CAMBRIDGE LIBRARY COLLECTION

Books of enduring scholarly value

Physical Sciences

From ancient times, humans have tried to understand the workings of the world around them. The roots of modern physical science go back to the very earliest mechanical devices such as levers and rollers, the mixing of paints and dyes, and the importance of the heavenly bodies in early religious observance and navigation. The physical sciences as we know them today began to emerge as independent academic subjects during the early modern period, in the work of Newton and other 'natural philosophers', and numerous sub-disciplines developed during the centuries that followed. This part of the Cambridge Library Collection is devoted to landmark publications in this area which will be of interest to historians of science concerned with individual scientists, particular discoveries, and advances in scientific method, or with the establishment and development of scientific institutions around the world.

The Monster Telescopes, Erected by the Earl of Rosse, Parsonstown

William Parsons (1800–67), 3rd Earl of Rosse and later president of the Royal Society, was responsible for building the largest telescope of his time, nicknamed the 'Leviathan'. It enabled the Earl to describe the spiral structure of galaxies. This volume reissues two contemporary accounts of the telescope. The first, by the Earl himself and published in 1844, provides a comprehensive description of the workings of both the 'Leviathan' and the smaller telescope which preceded it, together with detailed accounts of the construction of both telescopes. The second, by an anonymous author, first appeared in the Dublin Review in March 1845. Together with a brief description of the telescopes, it outlines the history and problems of telescope manufacture from Galileo onwards. This situates the telescopes, and the difficulties the Earl faced during the 18 years he took to build the 'Leviathan', in their wider context.

Cambridge University Press has long been a pioneer in the reissuing of out-of-print titles from its own backlist, producing digital reprints of books that are still sought after by scholars and students but could not be reprinted economically using traditional technology. The Cambridge Library Collection extends this activity to a wider range of books which are still of importance to researchers and professionals, either for the source material they contain, or as landmarks in the history of their academic discipline.

Drawing from the world-renowned collections in the Cambridge University Library, and guided by the advice of experts in each subject area, Cambridge University Press is using state-of-the-art scanning machines in its own Printing House to capture the content of each book selected for inclusion. The files are processed to give a consistently clear, crisp image, and the books finished to the high quality standard for which the Press is recognised around the world. The latest print-on-demand technology ensures that the books will remain available indefinitely, and that orders for single or multiple copies can quickly be supplied.

The Cambridge Library Collection will bring back to life books of enduring scholarly value (including out-of-copyright works originally issued by other publishers) across a wide range of disciplines in the humanities and social sciences and in science and technology.

The Monster Telescopes, Erected by the Earl of Rosse, Parsonstown

With an Account of the Manufacture of the Specula, and Full Descriptions of All the Machinery Connected with These Instruments

WILLIAM PARSONS

CAMBRIDGE
UNIVERSITY PRESS

CAMBRIDGE UNIVERSITY PRESS

Cambridge, New York, Melbourne, Madrid, Cape Town, Singapore,
São Paolo, Delhi, Dubai, Tokyo

Published in the United States of America by Cambridge University Press, New York

www.cambridge.org
Information on this title: www.cambridge.org/9781108013758

© in this compilation Cambridge University Press 2010

This edition first published 1844
This digitally printed version 2010

ISBN 978-1-108-01375-8 Paperback

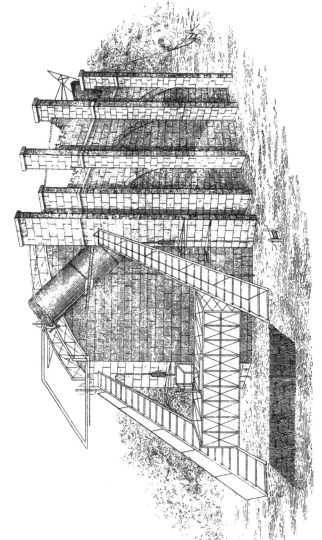

The Monster Telescope

THE

MONSTER TELESCOPES,

ERECTED BY

THE EARL OF ROSSE, PARSONSTOWN,

WITH AN ACCOUNT OF

THE MANUFACTURE OF THE SPECULA,

AND FULL DESCRIPTIONS OF ALL

THE MACHINERY

CONNECTED WITH THESE INSTRUMENTS.

Illustrated by Engravings.

SECOND EDITION.

PARSONSTOWN
SHEILDS AND SON, CUMBERLAND-SQUARE.
LONDON: DUNCAN AND MALCOLM, PATERNOSTER-ROW.
DUBLIN: JOHN CUMMING, AND W. CURRY.
MDCCCXLIV.

PREFACE.

To satisfy a curiosity that is naturally excited by any thing new, great or uncommon, I tried to obtain all the information I could, connected with the Telescopes I endeavour to describe. To gratify a similar feeling in others, more remote from my opportunities of looking on, I venture to publish an account of what I have seen.

As I am hardy enough to do so without any assistance from, or even the cognizance of the noble projector of those instruments, whose liberality in diffusing his knowledge and wish for its promotion, leave me no uneasiness on this point, so I do not expect to give that information which men of deep research or mathematically close enquiry would desire. There are some particulars which might, perhaps, be more enlarged upon with advantage, but it has been my aim to place before the general reader such an account as will make the manufacture of the Specula, and the mechanism of the Telescopes, as plainly understood as could be expected, without entering with too much tediousness into minute details. I have been as explicit as possible in the history of the compound three foot Speculum, knowing that individuals whose inclination would lead them to construct Specula on a large scale, without possessing the pecuniary advantages of Lord Rosse, will be naturally led to adopt a course the most manageable and economical, and one which does not appear to be the less certain of success. I have added illustrations of every part which is at all complicated, and where I thought they could be likely to assist the text, and trust that the endeavour I have made to enable my subject to be understood, will satisfy the desire of the public to understand it.

THE AUTHOR.

Parsonstown, September 28, 1844.

PREFACE TO SECOND EDITION.

The rapid sale of the First Edition, together with the many very favorable opinions expressed in reference to it by the Newspaper Press, leads the Author to hope that this work has been presented in an acceptable form. In the present edition he has therefore merely supplied a few trifling omissions which an unavoidable hurry in bringing out the former, occasioned.

Parsonstown, November, 1844.

CONTENTS.

PART I.

PART II.

THE MONSTER TELESCOPES.

PART I.

Lord Rosse having satisfied himself by experiments with lenses that the Refracting Telescope could not be much improved, turned his attention to Reflectors; and as the first object of experimenters had always been to increase the magnifying power and light by the construction of as large a mirror as possible, so was it to this point that his Lordship's attention was also directed.

Previous to his experiments, there had not been any instrument constructed, with the exception of Sir W. Herschell's, which had given an opportunity of sufficiently well enjoying the advantage of the Reflecting Telescope; and even of this it has been lately stated,* that it possessed but little, if any, practical superiority over others of smaller size.

*Vide Dr. Robinson's speech at the Cork meeting of the British Association, 1843.

Since Newton manufactured his Specula until the present day, there have been several opinions both as to the metals to be employed in their construction, and the quantities in which they should be mixed—some have recommended various proportions of tin and copper; some have added arsenic; some silver; some antimony; and others, the three together—the general aim of all the operators being to increase the whiteness, and to diminish the porosity and brittleness of the compound; for the last named property has destroyed a much greater number of Specula than it has allowed to be completed—it is the "asses' bridge" over which few have been fortunate enough to travel. The compound metal is exceedingly brittle, and requires the most cautious and well regulated cooling after having been cast to ensure its ultimate success; the very operation of grinding—by the heat the friction produces—having often destroyed the labour of many weeks. The grinding of Specula used to be performed by the hand, no machinery having been deemed sufficiently exact; the tool on which they were shaped being turned to the required form, and covered with coarse emery and water, they were ground on it to the necessary figure,

3

and afterwards polished by means of putty or oxide of tin, spread on pitch, as a covering to the same tool in the place of the emery. To grind a Speculum of six or eight inches in diameter was a work of no ordinary labour; and such a one used to be considered of great size. Mr. Ramage's Reflecting Telescope—the mirror of which is fifteen inches in diameter, with a focal distance of twenty-five feet, was until lately, the largest in use. Sir J. Herschell's, which stands in the place of his father's large one, has a Speculum eighteen inches in diameter, and a focal distance of twenty feet. With the difficulties of the undertaking, and the small success of his predecessors, Lord Rosse was fully acquainted, but he set about the work with a zeal that was a presage of his triumph. It is perhaps unnecessary to state how fitted his Lordship is in every way to accomplish a work which requires the combination of so many qualities—talent to devise—patience to bear disappointment—perseverance—profound mathematical knowledge—mechanical skill—uninterupted leisure from other pursuits; and yet these would not have been sufficient if a great command of money had not been added. Fortunately the world has seen them all combined, and their application to the highest branch of scientific enquiry. The fact of the great Telescope alone having cost certainly not less than twelve thousand pounds, shows how few individuals could have successfully brought so large an instrument to a happy conclusion; and if to this be added the money expended on the Telescope with the Speculum three feet in diameter, and the various sums laid out in experimenting, we will see what a splendid offering has been made at the shrine of science by one of her most devoted admirers.

After many trials as to what combination of metals was most useful for Specula, both as regarded whiteness, porosity and hardness, Lord Rosse found that copper and tin, united very nearly in their atomic proportions, viz:—copper, 126·4 parts to tin 58·9 parts, was the best. This compound, which is of admirable lustre and hardness, and has a specific gravity of 8·8, he has used with both the small and large Specula, and he finds it to preserve its lustre with more splendor, and to be more free from pores than any other with which he is acquainted. Having ascertained the proportion to be used, he set about the casting of the

Speculum. The difficulties attending this process were such, that instead of having the Reflector, which was to be three feet in diameter, in one piece, he tried the expedient of casting it in sixteen separate portions; the mirror when entire, presenting the form and appearance of figure 1. When cast,

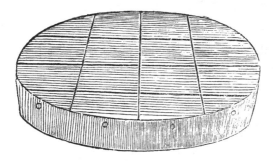

Fig. 1.

the pieces were fixed on a bed composed of zinc and copper, —mixed in the proportion 2·75 of the latter to 1 of the former—a species of brass, which expanded in the same degree by heat as the pieces of the Speculum themselves; they were then ground as one body to a true surface, and when polished were found to answer remarkably well. The particulars in the manufacture of this compound Speculum requiring attention are, First—The method employed to find the quantity of zinc and copper which would when mixed expand in the same degree as the Speculum itself. Secondly —The form into which the brass was made. Thirdly—The manner in which the Speculum was joined to the bed. Fourthly—The method in which the pieces were cast. And Fifthly—the grinding and polishing.

In order to find the proportion of the zinc and copper, a bar of the Speculum metal fifteen inches long and one and a quarter thick, was joined accurately to a bar of the brass of the same length, but only three-quarters of an inch thick; they were then placed in a vessel of water, whose temperature was lowered to 32^0 F. and a fine hair line drawn evenly and continuously across both; the temperature of the water was then raised to 212^0 F. and during the expansion, the line was examined with a microscope ; when it remained perfectly strait across both Speculum metal and brass, it was evident that they had equally expanded, and that the desired compound had been attained. There was much difficulty experienced in the melting of the brass, as the zinc being volatile it was in less quantity after this operation than before it, and the loss it sustained was not always the same. After many endeavours to remedy this defect, it was found that a certain result was obtained by spreading over the metal a layer of powdered charcoal two inches in thickness; the loss sustained was always in this case constant, and amounted to $\frac{1}{180}$th of the zinc employed.

The bed was cast in eight pieces, each being of the shape of figure 2, which is seen in reverse. The depth of the bed

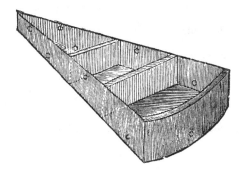

Fig. 2.

was five inches and a-half, and it weighed four hundred and fifty pounds. The eight portions were soldered together along the sides, and iron bolts used to secure the joinings more perfectly. However, it was found when the Speculum was fastened on that the joints had given in different places, as was evident by the effects produced in using the Telescope. The Speculum was therefore taken off, and the following device, technically termed "burning," was had recourse to. The entire brass work was imbedded in sand and covered to the depth of three inches. Down to the centre was bored a hole, one inch and a-half in diameter, and into this from a height of ten inches was poured a stream of melted brass, similar in composition to that of which the bed was made, until forty-five pounds had overflowed the surface of the sand. This operation was repeated in thirty-four different places on the combined edges of the portions of the brass bed. The constant heat kept up by the stream on these parts fused the metal, which contracted on cooling, and united them perfectly in these situations. The other surface of the bed was attempted to be managed in the same way, but the process failed here; some of the brass was therefore chiselled out of the necessary points and more melted in, which contracted when cool and fully answered the desired end.

When it was found that the joinings were now secure, the Speculum metal and bed were again soldered together;

this, however, was no easy matter, and gave rise at first to much annoyance and loss of time. It was attempted to solder them by tinning the brass bed, and placing the Speculum on this with resin interposed, then raising the temperature and using slight pressure when the tin and resin were fused. However, the latter becoming decomposed during the elevation of the temperature effectually prevented any union taking place. Sal Ammoniac was tried with no better success. The process was therefore modified, and the junction of the metals was accomplished in the following manner.

The pieces of the Speculum metal being ground by the hand to fit the brass bed, they were scraped on one side with hard steel chisels, no spot being left untouched and made perfectly dry; the brass was also scraped and tinned, and all the resin washed off with turpentine, and afterwards with soap and water. The Speculum was then laid on the brass, and both were placed in an oven with an iron bottom and supported on bricks. When the tin was fused, which was effected in eight hours by a gradually raised temperature, melted resin and tin were poured in between the plates,

B

and they were moved gently backwards and forwards. The fire was withdrawn when the tin commenced acting on the Speculum, and the pieces being settled into the places where they were to remain—$\frac{1}{20}$th of an inch apart—all was allowed to cool gradually, and in five days it was ready to be ground.

The casting of a Speculum, and its preservation from breaking when cooling, has always been found a matter of the greatest difficulty. The compound is so exceedingly brittle that any sudden change of temperature is likely to destroy it; and it requires to be so free from pores that unless great care is taken to give exit to all the air contained in the melted metal before setting, it is rendered useless. In casting the pieces of the compound Speculum, Lord Rosse first tried the plan recommended by Mr. Edwards, making the mould in sand, but the Speculum was almost always full of flaws, and generally flew in pieces when setting; on attempting to put the parts together again, it was found that they no longer fitted in their places. It was evident from this circumstance that they must have undergone some great strain; and it was judged that this had been produced by the unequal cooling of the different portions

—the edges becoming solid first left the centre fluid, and this being, when about to set, unable to contract, was strained when no longer ductile. The failures in this way were so frequent that another plan was tried, which was this:—A number of equidistant plates of iron were placed in a crucible with the fluid Speculum metal, under the idea that it would be divided when solid, into an equal number of Specula. But this plan had no success at all; the plates were always full of flaws.

The next attempt was much more encouraging; a block of the Speculum metal was cut into plates by a circular sawing machine, the blades of which were of soft iron, kept constantly wetted with emery and water. This process gave pieces that were entirely free from flaws; however, the texture was not perfectly uniform throughout, for near the circumference the arrangement of the particles was not just the same as in the other parts, so that it was feared that a good Speculum would not be formed. It has, however, never been sufficiently well tried, another method having been discovered before this was fairly tested. This process which has deservedly stamped Lord Rosse's

name with celebrity, and reduced the casting of Specula
to a certainty, is perhaps the most deserving of praise
of all his Lordship's works. The simplicity of the con-
trivance, probably at first sight, makes it appear a result of
no great wonder: but like the plan pursued by Columbus
to make the egg stand, it is only easy when known. It
has been stated that the chief conditions so hard to be
attained in the manufacture of Specula are, first—certainty of not breaking in cooling; and secondly—a freedom from pores, if it has been so fortunate as to escape
the first. Now we saw that the mould of sand did not
fulfil the first intention, from the unequal contraction the
metal suffered when cooling; on this account a mould
was made of cast-iron, in the hope that its high conducting power would lower the temperature of the bottom
portions of the fluid metal first, and abstracting the heat
much more quickly than it could escape from the surface,
it would cool the metal gradually towards the top, and
in this manner, prevent any part of it being strained.
In order to keep the under surface of the mould at a low
temperature, a jet of cold water was made to play on it

while the Speculum metal was cooling; but unfortunately, the cold water almost always cracked the mould, and of course, the casting was destroyed when this occurred before solidity had taken place. This plan was, therefore, also unsuccessful. It was then attempted to procure a perfect casting by making a mould having the sides of sand and the bottom of iron, and using no jet of cold water. This was so far successful that it answered the purpose for which it was intended; that is, cooling the metal gradually from the bottom and preventing any flaws or breaking; but the air, not having the porous sand beneath to give it exit, lodged between the mould and the iron bottom, and formed large holes which rendered it useless. Now the great object to be attained was to find some way of allowing the air to escape, but still to retain the iron bottom. In effecting this consists the great improvement in casting Specula, without which, the magnificent instruments Lord Rosse now possesses would never have been in existence.

Instead of the bottom of the mould being of solid cast iron, it was made by binding together tightly layers of hoop-iron, and turning the required shape on them edgewise. This mould fulfilled all the requisite intentions. The iron shape conducted the heat away through the bottom, and cooled the metal towards the top in infinitely small layers, while the interstices, though close enough to prevent the metal escaping, were sufficiently open to allow the air to penetrate. The next casting was perfect, and in all succeeding trials the results have continued to be successful. The shape requires to be perfectly dry before being used, and is heated to 212⁰ F. without impairing its cooling power. It must be of sufficient thickness to conduct rapidly; never being thinner than the Speculum to be cast. The metal entered the mould by the side as rapidly as possible; and before being cast was heated, until when stirred with a wooden pole, the carbon reduced the oxide on the top, and produced a clear and brilliant surface. Immediately on becoming solid, it was removed to an annealing oven, and cooled slowly. A plate nine inches in diameter, requires three or four days to anneal. We will here notice a circumstance worthy the attention of those who may be in future engaged in a similar attempt to his Lordship's. In melting the metal it was found that when fluid, it oozed through the bottom of the cast iron crucible employed to contain it. The imperfect part was cut out, and the hole stopped by a wrought iron screw. This, however, was ineffectual, as other parts of the bottom were alike defective; and it having been discovered that the porosity of the crucible resulted from the fact of its being cast with the face downwards, which caused the small bubbles of air to lodge in the then highest part, Lord Rosse had some cast in a contrary direction in his own laboratory, and obtained vessels perfectly secure and staunch. In the melting of the metal, turf fires were found sufficiently good, without being strong enough to endanger the crucibles, and they were more steady than those produced by coke.

The process by which the plates were soldered to the brass bed has been already described; and to complete the history of the compound Speculum, we have now to notice the grinding and polishing. Before this, however, we may remark that the porous iron bed above described, gave in so many trials such perfect results, that a Speculum, three feet in diameter, was cast in one piece, and went through all the steps of casting, cooling, &c. with success. There was no difference in the manipulation of this and the compound Speculum, except, of course, that it did not require any brass bed to be soldered to; it was cast in an open mould and was allowed time in proportion to its size to anneal. It is this three foot Speculum that is at present in the Telescope erected in his Lordship's lawn; the machinery of which will be described further on. It was not however, chosen for this situation on account of its superiority over the compound Speculum, as there could be scarcely any difference found to exist between them: there is, however, a slight diffraction of light, caused by the edges of the pieces of the latter, but it does not interfere to such a degree with its performance as to be taken into account. The processes of grinding and polishing were precisely similar in the case of each, and the machinery was the same. We premise that the moving power employed was a steam engine of three horse power, which was also made use of for many of the other processes connected with the manufacture: one horse power is sufficient for grinding. The several parts of the ma-

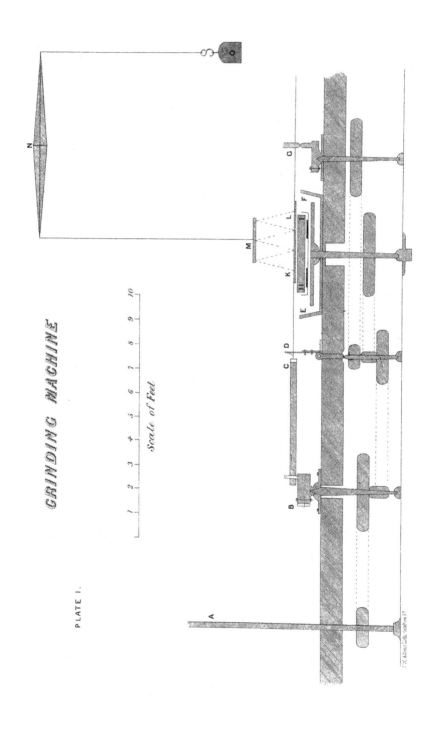

GRINDING MACHINE

PLATE I.

Scale of Feet

chinery are connected by bands of leather in place of toothed wheels, which would be more liable to accident.

Specula used, before the existence of these we are describing, to be polished with the hand, and were subject to the greatest uncertainty of result. The shape on which they were worked was placed on a support, so that the manipulator might move round it. The Speculum was then ground down by rubbing it backward and forward, and at the same time, getting a sort of circular motion from the movements of the operator round the support. The process was laborious and not to be depended on. The velocity with which the hand moved varied constantly—the temperature was never regulated, and the pressure on the several parts of the surface was not the same.

In order to get rid of these defects, and if possible, to produce a surer and better result, Lord Rosse invented his machine for grinding. A sketch of this is seen in plate 1, copied from the Philosophical Transactions for 1840.—A is a shaft connected with the Steam Engine; B an excentric adjustable by a screw-bolt to any given

C

length, from 0 to eighteen inches; C a joint; D a guide; E F a cistern for water, in which the Speculum revolves; G another excentric adjustable, like the first, to any length of stroke, from 0 to eighteen inches. The bar, D G, passes through a slit, and therefore the pin at G necessarily turns on its axis at the same time as the excentric. H I is the Speculum in its box, immersed in water to within an inch of its surface, and K L the polisher which is of cast iron, and weighs about two hundred and a-half. M is a round disk of wood connected with the polisher by strings hooked to it in six places, each two-third of the radius from its centre. At M there is a swivel and hook to which a rope is attached, connecting the whole with the lever, N, so that the polisher presses on the Speculum with a force equal to the difference between its own weight and that of the counterpoise, O.

The polisher is connected with the machinery by means of a large ring of iron, which loosely encircles it, ending in the bars which run through G and D. The wheel work underneath the table shows the manner in which the polishing goes on. Instead of either the Speculum or the polisher

being stationary, both move with a regulated speed ; the ring
of the polisher, and therefore the polisher itself, has a transverse and a longitudinal motion; it makes eighty strokes in
the minute, and twenty-four strokes and a-half backward
and forward for every revolution of the mirror, and at the
same time, $1\frac{72}{100}$ strokes in the transverse direction. The extent of the latter is $\frac{27}{100}$ of the diameter of the Speculum.
The polisher has another motion independent of the ring, for
at the turn of the excentric, being for a little time free, it is
carried for a short distance round, lying on the Speculum.
In this way it makes one revolution for every fifteen of the
mirror. The peculiarities of this mechanism are, 1st. the
mechanism itself, as before this was used it was not thought
possible to grind by machinery,—2nd. placing the Speculum
with the face upward,—3rd. regulating the temperature by
having it immersed in water, which is usually at 55^0 F.
—4th. regulating the pressure and velocity. The pressure allowed to be exerted on a three-foot Speculum is
ten pounds. The improvements effected in grinding and
polishing mirrors by this machine are wonderfully great;
and like all the other improvements invented by Lord

Rosse, are the result of close calculation and observation. We are indebted to his powers of reasoning for all his works, and have to thank the blind goddess, Chance, for none. She is indeed so little enamoured of his Lordship that when she visits him at all, which is seldom, she appears in the shape of failure.

The following quotation from Lord Rosse's publication in the Philosophical Transactions, will enable us to understand the theory of the Machine:—"Having observed that when the extent of the motions of the polishing machine was in certain proportions to the diameter of the Speculum, its focal length gradually and regularly increased, that fact suggested another mode of working an approximate parabolic figure. If we suppose a spherical surface, under the operation of grinding and polishing, gradually to change into one of longer radius, it is very evident that during that change, at no one instant of time will it be actually spherical and the abrasion of the metal will be more rapid at each point as it is more distant from the centre of the *face*. When, however, the focal length neither increases nor diminishes, the abrasion will become uniform over the whole surface producing a spherical figure. According, however, as the focal length (the actual average amount of abrasion during a given time being given) increases more or less rapidly, the nature of the curve will vary, and we might conceive it possible, having it in our power completely to control the rate at which the focal length increases, so to proportion the rate of that increase, as to produce a surface approximating to that of the paraboloid. Of course the chances against obtaining an exact paraboloid are infinitely great, as an infinite number of curves may pass between the parabola and its circle of curvature, and it is vain to look for a guide in searching for the proper one in calculations founded on the principles of exact science, as the effect of friction in polishing is not conformable to any known law; still from a number of experiments it might be possible to deduce an empirical formula practically valuable: this I have endeavoured to accomplish."

The grinding of the Speculum to the proper figure then depends on the relative velocities of the different parts, as before stated. The substance made use of to wear down the surface was emery and water; a constant supply

of these was kept between the grinder and the Speculum.
The Grinder is made of cast iron, with grooves cut length-
ways, across, and circularly, on its face; there are twenty-
five grooves, crossed by as many more, which are quarter
of an inch wide, and half an inch deep   The circular
grooves, of which there are thirteen concentric with the
polisher, are three-eighths of an inch deep, and quarter of an
inch wide.   The polisher and speculum have a mutual action
on each other; in a few hours, by the help of the emery and
water, they are both ground truly circular, whatever might
have been their previous defects.   The grinding is continued
until the required form of surface is produced, and this is
ascertained in the following manner.   There is a high tower
over the house in which the Speculum is ground, on the
top of this is fixed a pole, to which is attached the dial
of a watch; there are trap doors which open, and by
means of a temporary eye-piece at the calculated distance, allow the figure of the dial to be seen in the
Speculum brought to a slight polish.   If the dots on the
dial are not sufficiently well defined the grinding is continued; but if it works satisfactorily, the polishing is

23

commenced. This process, like the others, was in the beginning attended with much difficulty and annoyance. To polish Specula, the same tool which grinds them is covered with a layer of pitch, and on this is spread either putty or oxide of tin, as used by Newton, or oxide of iron, commonly called rouge, as used by Lord Rosse.— Now, in trying to polish the three foot Speculum it was found that the pitch, which should be spread in a thin layer and very evenly over the surface, collected the abraded matter in some places more than in others, and of course, lost its required shape. If the pitch expanded laterally, so as to fit itself again to the Speculum, all would have gone on well; but, unfortunately, this could not take place unless the layer was made a great deal thicker than it could be satisfactorily used. In polishing small mirrors a thin layer could expand sufficiently laterally, on account of the small extent of its surface; but experience proved that the thickness of the layer should increase with the size of the Speculum, until, setting out from the depth of half-a-crown, which is requisite for a very small mirror, it would come to an unmanageable size before it reached that sufficient for one three feet in diameter. In order to get over this difficulty, and make room for the pitch to expand without increasing its thickness, it was cut into grooves along the surface with decided advantage, but the furrows filled up in a short time again, and it became, of course, as it was in the beginning, and the same impediment occurred. When this failed it was thought that if the polisher itself was cut into grooves it must have the desired effect, and the result proved the conjecture to be correct. The pitch might then be reduced to a very thin layer, being in the condition of a number of very small polishers united together, and the lateral expansion being allowed for, without danger of filling up the grooves, the polishing proceeded in a most satisfactory manner. However, it was not yet perfect. It requires for the production of a true surface, that the pitch should always be in general contact with the Speculum, and it should be as hard as possible, consistent with allowing the rouge to imbed itself in it. The first condition is easily satisfied by the grooves above described, provided the pitch be

soft enough to expand; but if this be the case, it would
be too soft for producing the true surface; therefore, to allow
of the expansion, and at the same time have the pitch very
hard, Lord Rosse first spread over the tool a layer of soft
and then one of hard resin—which he preferred to pitch on
account of the gritty impurities of the latter—and combined
both qualities in one. The manner in which the resin was
prepared to be substituted for pitch was this: a large quantity was melted, and spirits of turpentine about one-fifth of
its weight, was poured in, and both well mixed together;
—an iron rod dipped into this mixture brought up a little
and allowed its temper to be tried;—after being cooled
in water at 55⁰ F. if the thumb nail made a slight but
well marked impression the materials were mixed in proper
proportions. This was then divided into two parts so as to
make the hard and the soft layer. For the first, quarter the
quantity of wheaten flour was mixed with one part, and
boiled until the water of the flour was expelled, and the
mixture became clear—some of the turpentine was also driven
off. There was then added an equal quantity of resin. This
was sufficiently hard for the upper layer. The flour increases

D

its tenacity, and diminishes its adhesiveness, and so prevents the abraded matter running loosely between the polisher and the Speculum; the under layer was only the mixed
turpentine and resin; each layer is $\frac{1}{25}$ of an inch thick; the
first is laid on with a brush, the polisher being previously
heated to 150° F.; when the temperature is lowered to 100° F.
the second layer is laid on in the same way, and when the
temperature falls to 80° F. the polisher is placed on the Speculum whose shape it takes, the warm resin accommodating itself to its figure. Were it not for the bad conducting power
of the resin it would be a dangerous experiment to lay the
polisher heated to 80° F. on the Speculum, as it would most
probably crack it in pieces, but when the resin is interposed
it can be done with safety.

When the polisher was thus prepared it was smeared over
with rouge and water, and a supply of the same being kept up,
it was by the machinery which was employed for grinding, and
with the same relative velocities of stroke, &c. quickly brought
to a fine black polish. The friction does not alter the form of the
pitch by the heat it produces, as the Speculum is kept floating
in water, whose temperature is 55° F., the same temperature

PLATE II.

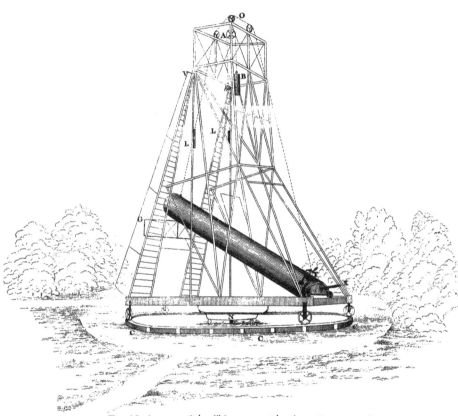

The Machinery of the Telescope with the 3 Foot Speculum

Allen's Litho 16 Grafton St

to which the resin was regulated at first. The length of time required for polishing was six hours. When this was completed it was supported on an equilibrium bed; that is, three iron plates, each being one-third of a circular area, which joined together made up a disc of the same size as the Speculum itself; these rest in the Speculum box on points at their centres of gravity, so that no flexure of the wood can affect the mirror. Plate 2 is a sketch of the machinery by which the Telescope is suspended and worked. The tube is twenty-six feet long;—the focal distance of the Speculum being twenty-seven—the whole machinery is supported on four wheels, which run on the iron circle, C C C. This circle is thirty feet in diameter, and is graduated and marked with the cardinal points. The pivot on which the machinery turns passes through a beam of wood, N, which is fastened to the under part of the large cross beam. It is about four inches from the ground; through one end run two screws, which by being turned have, when pressed downwards, a tendency to raise the beam at that end; there is a lever imbedded in the earth with its long arm loaded, the short one pressing up against the other end of the beam, N, so that the whole machinery is counterpoised to any required extent, and the weight on the wheels which run on the circle is reduced to such a degree that the instrument can be turned round to any position with the greatest ease. Across the end of the tube is fixed an axle, P, to which a small wheel is attached on either side, these run along a railway when the tube is elevated or depressed. The tube itself is counterpoised by the weight, B, hanging over the pulley, A. In order to raise or depress it, a small wheel and axle, T, is fixed to one end, a rope runs from this over the top wheel, O, and is joined to the other end; when the rope is wound round the axle, of course, the free end of the Telescope is raised. Instead, however, of the rope ending at the free extremity by being fixed immediately to the tube, it meets and is fastened to a smaller rope which runs through a pulley on the end of it; one extremity of this small rope is fastened to the tube, and is as it were a continuation of the large one; the other extremity is wound round a small wheel and axle which is situated near the eye-piece, so that the observer is able, by shortening or lengthening this, to raise or depress the tube for short distances. The gallery, G, is supported on the step of the ladder by a catch, and is counterpoised by the weights, L L, hanging over the pulleys, V V. It is of course, raised and lowered with the tube. X is a lever attached to the Speculum box for opening it: a vessel of lime is kept in the box for absorbing all the moisture which might tarnish the mirror. The working of this instrument is most satisfactory; the double stars are delineated with great brilliancy and some of the Nebulœ have been resolved in a more perfect manner than had previously been accomplished by any other Telescope. The great difference of appearance which these Nebulœ present when viewed in this instrument compared with the published plates of Sir J. Herschell is quite conclusive as to its great superiority. It certainly was the finest instrument in the world until it was surpassed and even thrown into comparative insignificance by the Speculum, six feet in diameter, which has arrived at the bounds beyond which the laws of matter forbid human ingenuity to pass, and marks one spot in the circumference of that great circle that defines our powers, which is so seldom reached, and for that reason so little known.

THE MONSTER TELESCOPES.

PART II.

In the account of the three-foot Speculum which Lord
Rosse published in the Philosophical Transactions for 1840,
he speaks of the possibility of one six feet in diameter being
cast. It might at that time have been considered as little
less than a chimera by those who were not sufficiently acquainted with the experiments that had been made in his
Lordship's laboratory, and there were not wanting some who
denied altogether the practicability of the design. Various
reasons were given why the attempt should be a failure, and
many calculations entered into to prove the little benefit to
be derived even supposing a perfect casting were obtained.—
But fortunately others thought differently; the idea had no

sooner occurred to Lord Rosse than he determined to put it
to the test, and we may say, without flattery, that no absur-
dity was likely to occupy a mind like his. The attempt has
been made, and the result is perfect success.

As yet we cannot say how far it may advance our knowledge of the celestial spheres, or help us to understand
more fully the mechanism of the universe; but this at
all events is certain, that be the advantage great or small,
it is the last step that can be taken to enlarge our ac-
quaintance with those distant bodies, and all that is ever
likely to be brought before us will now be seen. But although we cannot hope to know more than will be revealed
to us by this instrument, we may anticipate a great addition to our present knowledge, and perhaps it may be the
means of rendering us as familiar with the secrets of the
starry heavens as our finite intellects are capable of un-
derstanding.

To describe the processes by which this six foot Speculum
was manufactured, would be repeating nearly what we have
already said of that three feet in diameter. The composition of the metal and the manipulation of the casting,

grinding, polishing and annealing were the same, except of course on a larger scale, and the only alteration which took place was in consequence of its greater size. The foundry however in which it was manufactured was constructed expressly for the purpose. There was a chimney built from the ground, six yards high, and five and a-half feet square at the base, tapering to four feet at the top: it was rectangular. At each of three of its sides communicating with it by a flue, was sunk a furnace eight feet deep, (not including the ashpit, which was two,) and five and a-half feet square, with a circular opening four feet in diameter; they were lined with brick. About seven feet from the chimney was erected a large crane with the necessary tackle for elevating and carrying the crucibles from the furnace to the mould, which was placed in a line with the chimney and crane, and had three iron baskets supported on pivots hung round it. Four feet further on in the same line was the annealing oven— Figure 3 is a ground plan of the whole,—1 is the chimney; 2 2 2 the furnaces; 3 the crane; 4 4 4 the iron baskets; 5 the mould, and 6 the oven. The crucibles which contained

E

the metal, were each two feet in diameter, two and a-half feet deep, and together weighed one ton and a-half; they were of cast-iron, and made to fit the baskets at

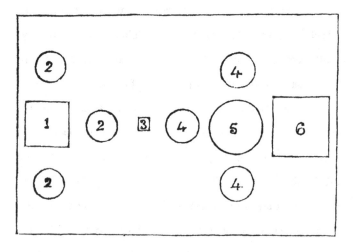

Fig. 3.

the side of the mould; these baskets were hung on wooden uprights on two pivots; to one of these pivots was attached a lever, by depressing which the basket might be turned over, and the contents of the crucible poured into the mould. It has already been described how the bottom of the mould is made, by packing together layers of hoop-iron, on which the certainty of the safety of the

Speculum depends. This bottom, which is six feet in diameter, and five inches and a-half thick, was made perfectly horizontal by means of spirit levels, and was surrounded by a wooden frame; a wooden pattern, the exact size of the Speculum, being placed on the iron, sand was well packed between it and the frame and the pattern was removed. Each of the crucibles containing the melted metal was then placed in its basket, and every thing being ready for discharging their contents, they were at the same instant turned over, and the mould being filled, the metal in a short time safely set into the required figure. While it was red hot and scarcely solid, the frame work was removed, and an iron ring connected with a bar which passed through the oven, being placed round it, it was drawn in by means of a capstan at the other side on a railroad, when, charcoal being lighted in the oven and turf fires underneath it, all the openings were built up and it was left for sixteen weeks to anneal. It was cast on the 13th of April, 1842, at 9 o'Clock in the evening. The crucibles were ten hours heating in the furnaces before the metal was introduced, which in about ten hours more was sufficiently fluid to be poured. When the oven was opened the Speculum was found as perfect as when it entered it. It was then removed to the grinding machine, where it underwent that process, and afterwards was polished without any accident having occurred. It weighs three tons, and lost about one-eighth of an inch in grinding. Lord Rosse has since cast another Speculum of the same diameter, four tons in weight. He can now, with perfect confidence, undertake any casting, so great an improvement has the form of mould which he has invented proved. The Speculum is placed on an equilibrium bed in the same manner as the smaller one; this bed is, however, composed of nine triangular pieces instead of three, resting on points at their centres of gravity; on each corner of these nine pieces is another also resting on its centre of gravity,—in all twenty-seven pieces,—presenting a flat surface, on which the Speculum lies, and which, from its construction, prevents any change in the box or tube produced by warping or other cause affecting the mirror: the pieces were lined with pitch and felt before the Speculum was placed on them. The bad conducting power of these substances is intended as a security against any variation of temperature being suddenly conveyed to the Speculum. A vessel of lime is kept in connexion with the Speculum box to absorb the moisture, which otherwise might injure the mirror.

The Frontispiece is a perspective view of the machinery by which the great Telescope is worked. The tube is fifty-six feet long, including the Speculum box, and is made of deal one inch thick, hooped with iron. On the inside, at intervals of eight feet, there are rings of iron three inches in depth and one inch broad, for the purpose of strengthening the sides. The diameter of the tube is seven feet. It is fixed to mason work in the ground by a large universal hinge which allows it to turn in all directions. At each side of it at twelve feet distance, a wall is built which is seventy-two feet long, forty-eight feet high on the outer side, and fifty-six on the inner; the walls are thus twenty-four feet apart, and they lie exactly in the meridional line. The fixed end of the Telescope is in the centre of the enclosed space, and the free end turns round to either extremity, looking north or south as required. When directed to the south, the tube may be lowered till it becomes almost horizontal, but when pointed to the north it only falls until it is parallel with the earth's axis, pointing then to the pole of the heavens; a lower position would be useless, for as all celestial objects circumscribe that point, they will come into view above and about it. Its lateral movements take place only from wall to wall, and this commands a view for half-an-hour at each side of the meridian. With so large an instrument the most favourable circumstances must be combined, to allow of its being used with success; and as all bodies in the heavens are seen more perfectly when on or near the meridian, it was thought quite sufficient to have them for one hour in the field of view in their most manageable situation, and as they must also of course pass the meridian, nothing is lost by this limited range of the Telescope. There is a chain connected with that part of the tube which is uppermost when it points to the south, that runs over a pulley in a truss-beam at the northern end of the wall, and is wound round an axle on the ground. This elevates and turns the tube to the north. A beam

of wood, twenty-five feet long, is hinged at one end to
the mason work, which supports the large universal joint
on which the tube moves; this is loaded at the other end
by a weight, and from that is joined to the tube by a
chain thirty-five feet long. It is so managed that when
the tube has reached its perpendicular position, the weight
which is on a cross-beam is at its fullest extent from the
tube, and as the tube continues to move towards the north
this weight is raised, forming an angle with the horizon.
From each side of the tube runs a chain, which passes
round a pulley fastened to the wall, but which can turn on
a pivot to suit itself to the different situations of the tube;
the chain then runs under another pulley which is stationary, and ends by suspending a weight, which is thus a
counterpoise to the tube. This weight on either side is
also fastened to a chain, which hangs from a beam at
the northern extremity of the wall: when the tube is
pointed towards the south, the weights hang on the chains
which run from its sides over the pulleys, and so have
a tendency to elevate it; but as it reaches the perpendicular, these weights are prevented sinking down in a

straight line by the other chains which are fixed to the beam, and which, always continuing the same length, draw them out towards the extremity of the wall until they hang altogether on themselves, and exert no force at all on the chains which are connected with the tube: when this passes the perpendicular towards the north, the weights are again drawn back, and begin once more to counterpoise it.

Plate 3 will explain more clearly the different parts we have been attempting to describe. It is a view of the inside of the eastern wall, with all the machinery seen in section. A is the mason work in the ground; B the universal joint, which allows the tube to turn in all directions; C the speculum in its box; D the tube; E the eye-piece; F the moveable pulley; G the fixed one; H the chain from the side of the tube; I the chain from the beam; K the counterpoise; L the lever; M the chain connecting it with the tube; Z the chain which passes from the tube to the windlass over a pulley on a truss-beam, which runs from W to the same situation in the opposite wall—the pulley is not seen; X is a rail-road

PLATE 3

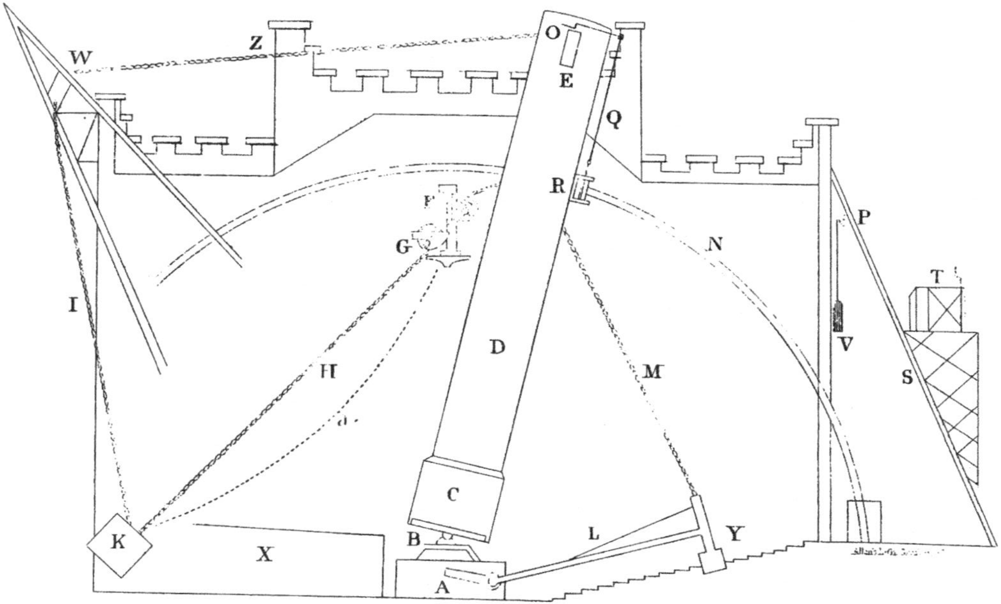

on which the Speculum is drawn either to or from its box—part is cut away to show the counterpoise. The dotted line, a, represents the course of the weight R as the tube rises or falls; it is a segment of a circle of which the chain I is the radius.

With a little attention to these several points, the working of the machinery, will, we think, be easily comprehended. We must first state that the Speculum is supported on three screws, which pass from the universal joint through the bottom of the box in which it lies, and is by this means so far unconnected with the tube, as to exert no pressure or strain upon it. The weight on the lever L sinks only fifteen feet under the horizontal position, it then rests on the ground, and is, of course, no load on the tube, which is when this happens 30⁰ F. above the horizon. Below this point the tube is sufficiently heavy to descend when the windlass unrolls the chain. Then suppose the tube makes the angle of 30° F. with the horizon, and that it is required to elevate it, the windlass is turned, and the chain being shortened, the desired effect is produced; but the labour of this would be immense if the counter-

F

poise K did not assist: this nearly balancing the tube, leaves but little exertion to be made at the windlass. However, the weight of the tube according as it ascends, is gradually becoming less and less, until it produces no strain at all on the windlass when it is quite upright. This must evidently be the case from the first principles of mechanics; for making the tube a lever, the length of its arm continually decreases as it approaches the perpendicular; therefore, if the counterpoise continued the same weight on the tube towards the end as it was in the commencement of the ascent, it would be too heavy and would keep it in its perpendicular position. In fact, the counterpoise must become lighter as gradually and as evenly as the tube itself, in order to continue to be just the same support to it all through its movement. The plan adopted to effect this is beautifully simple: a weight hanging freely in a perpendicular direction exerts its greatest force on the suspending point; if it be moved from the perpendicular, as much power as is required to effect this, is taken off from the same point; as will be evident to any person pushing aside a hanging body, he must supply a certain degree of force to keep it out of its perpendicular position; and this might be mathematically proved to amount to exactly the degree of weight that is taken off the point from which the body hangs. Now, from Plate 3 it will be at once seen how, when the tube is ascending and losing its weight, also lengthening the chain H, that on account of the chain I, whose length is always constant, the counterpoise K is moving from the perpendicular position under G, and, therefore, loosing its power on the tube, and approaching the perpendicular under W, and for this reason, transferring all its weight to the fixed chain I: when the tube passes the perpendicular, the chain H is again shortened, and the counterpoise begins once more to draw it back, so that the action of this tends to keep the tube always upright to whatever side it may point, and its power is always equal to the varying weight. Under these circumstances we see how easily and evenly the windlass can elevate the Telescope, and turn it to the north; but when it arrives there it must be brought back again; and this is accomplished by the lever L. As we have seen that the action of the tube and counterpoise is so regulated, that, in all positions the weights although always changing are equal to one another, so must the weight of the lever vary with its position in order to be a perfect balance on the tube; and this it evidently does. We said that when the tube was perpendicular, the weight on the lever is most effective; for it is at the farthest distance it can be from the support; it therefore, pulls down the tube when the windlass is unrolled; but we saw that the tube as it descends increases its weight; so that if the lever continued acting with the same power with which it commenced, the weight of both would be constantly increasing: this is prevented by the lever losing its force as it falls; for the weight thereby, of course, approaches the support and cannot be so active: but the approach to the support by its descent is so regulated to the increasing distance of the end of the tube in *its* descent by the chain M, that in the same degree as the latter gains weight the former loses it; and in this manner there is a constant equilibrium kept up between them. When the tube reaches within 30° of the horizon the lever rests on the ground, and the tube is thence able to descend by its own gravity. When the tube points to the north, the lever is elevated above the horizon, and has not, of course, so much power as when it coincided with it; but it is in this case helped by the counterpoise K, which always tends to bring the tube to the perpendicular. This continues to help it until it becomes itself sufficiently able, from its horizontal position, to do all the work; it then commences opposing it, but it now has the help of the increasing weight of the tube itself, and so all the parts are elegantly blended into one another with the most perfect concord and efficiency.

We have only yet described how the tube is elevated and depressed. The manner in which it is moved from side to side is this: there is a cast iron circle fixed to the eastern wall, marked N in Plate 3; from this nearly to the other wall there runs a wooden pole, about three inches by two in depth and breadth, through an iron bed R, which is fastened to the tube. On the face of this is a rachet in which plays a wheel connected with the rod Q. The handle O is terminated by an endless screw,

which works into a toothed wheel on the top of the rod
Q; therefore, it is evident, that by turning the handle O
the rod Q is also turned; this makes the wheel at R
work on the rachet, and as the end of this rachet is fixed
to the iron circle N, the tube must move either to or
from it, according to the direction in which the handle
is made to revolve. This is done with the greatest ease
by the observer. The end of the rachet connected with
the iron circle moves along it as the tube is elevated or
depressed. There is a universal joint on Q, which allows
the wheel and rachet to be used with great freedom, as
it prevents any stiffness in the rod. By these contrivances
the Telescope can be moved in any direction, and it is
admirable with what very little effort it can be adjusted
to any required place. It is kept quite firmly fixed in
any situation by the rachet and wheel—not the least tremor
being perceptible in its working. It is fitted up in a temporary way to be used as a transit instrument; for the piece of
wood which stretches from wall to wall, along which the
rachet runs, is graduated so as to mark right ascensions, and
there is fixed to the Speculum box a quadrant with a move-

able radius, along the upper surface of which is a spirit level; this being made horizontal, shows the angle at which the Telescope is placed and points out the declination. The scale for giving the right ascension is marked to lengths of a minute at the equator, which is 2-22 inches.

We will now show how the observer is to follow the tube, and be able to make use of it in all its different situations. The perspective view will enable the reader to comprehend the general plan. The gallery which stretches from ladder to ladder in front, is moved up and down with the Telescope; it is counterpoised by the weights which hang on the wall; a chain passes from each weight under ground, and being joined together, end at the windlass at the side of the wall. This raises and lowers the gallery, by means of which the observer follows the tube as far as the ladders allow him. In Plate 3 the gallery and box are seen in section. S is the gallery; T the box; V the counterpoise hanging over the pulley by the chain which runs under ground to meet that from the opposite side and be continued to the windlass. On a railway on the top of this gallery is the box in which the ob-

server is situated; one of the wheels on which it runs,
that on the right hand corner, is toothed on the side;
into this plays a smaller one, which is turned by a handle
which reaches as high as the top of the side of the box,
and this moves it from end to end of the gallery. The
ease with which the observer can wheel himself, and as
many as can stand in the box, is surprising. It requires no more effort than would be necessary to lift a
pound weight from the ground. However, the wheels being
thus regulated to need so small an exertion to produce
effect, of course lose in velocity what they gain in power:
but a slow motion is all that is required.

In order to keep up to the tube in its progress when it
passes the height to which the ladders rise, the galleries
on the western wall are constructed; they may be entered
either by a stairs on the outside of the wall, or directly
from the box on the ladder when this is at its greatest
height. Nothing would be simpler than to construct a
gallery extending along the top of the wall from end to
end, but the Telescope having to move twenty-four feet
from it, and the galleries having to follow, made the con-

sideration of the method to be adopted a matter of some moment. The plan pursued is this: there are three galleries, which, when joined together, make up a segment of the circle which the tube in its motion describes; therefore, by walking along this, the observer follows the tube from south to north. The middle gallery is twenty-six feet long, and the side ones twenty each. One is seen in Figure 4. A is the gallery; B is a paralelogram, formed

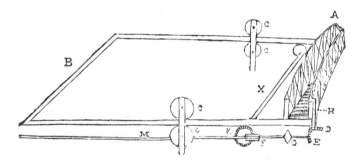

Fig. 4.

by wooden beams thirty-two feet long, more than sufficient to reach from wall to wall; on the top and bottom of the side beams is an iron rail for the purpose of allowing them to run easily between the wheels C C; when it is required to move the gallery, the handle H is turned;

G

this, by means of the wheel D working in E, turns the tooth wheel F, which makes K revolve. K is connected by the axle X, with a similar wheel at the opposite side, which is thus made revolve also. In each axle of the side galleries there are two universal joints, which allow the wheels to remain perpendicular while the axle is inclined; the pressure exerted on these wheels prevent them moving without at the same time carrying the galleries backward or forward according to the direction in which the handle turns. The wheels C C revolve as the beams pass between them. They are not connected with the handle, but are merely friction wheels, and they keep the gallery in its proper place, preventing it overbalancing when brought out from the wall. Before being used the galleries were wheeled out to their fullest extent, and loaded with a weight equal to that of twelve men,—which number each gallery can contain,—and were found perfectly secure and steady. The spring F on the rod M makes the galleries move with much more smoothness than they could do without it; when the handle is turned to bring them from the wall, the spring is first stretched and makes half a.

round, the gallery moving as far as the spring expands; the spring then closes, and brings on the iron rod with it through the wheel K; this is repeated, and the motion is scarcely felt, at the same time that it is quick enough for all its purposes; and the gallery when fully loaded might be moved by the hand of a child. The three galleries are managed in the same way, and answer their end exceedingly well. The magnifying power of the Speculum is so great, that objects pass from the field of view in a very short time, so that it is necessary to follow them by moving the Telescope in their direction; this is for the present accomplished by turning the handle O, plate 3; it allows them to be examined for some time. But there is intended to be machinery connected with the tube and galleries which will give an automaton movement to all, so that the Telescope may be used as an equitorial: as yet there is no appearance of this, but we believe, that before a twelvemonth it will be completed. There is no other machinery at present connected with the Telescope. All the purposes are fulfilled amply, by what has been already spoken of; the mechanism is

exceedingly simple, and uncomplicated, at the same time
that it combines the most perfect arrangement and effi-
cient working of the several parts. The entire is well
conceived; and the plan has been carried into effect in a
very substantial manner. All the works are of the strongest
and safest kind; every part that sustains any weight, or
to which any strain is applied, is at least, doubly equal
to its task, without appearing cumbrous or out of proportion. All the iron work was cast in Lord Rosse's
laboratory, by men instructed by himself; and every part
of the machinery of every description, was made under
his own eye, by the artizans in his own neighbourhood.
There did not a single accident, except one of a trifling
nature, occur during the whole proceeding; and the entire
design, just as it was planned before being commenced,
has been brought, without any material alteration, to a
safe and successful termination. The Telescope has been
used, and, although under unfavorable circumstances, it
promises extraordinary results. The first and most startling
proof of its superiority consists in the great quantity of
light that is present, even when using eye-glasses of very

high magnifying powers. The great difference in this respect between it and the three foot Speculum is most apparent: many appearances before invisible in the moon, have been seen, but of course, not yet examined. Lord Rosse hopes, we believe, to accomplish wonders with the Nebulæ and double stars. The Speculum has already exceeded his expectations; and there is not, we are sure, a single individual who will not be delighted to hear so; not so much at first for the great advance it will give to Astronomy, and the triumph it secures for Science and mechanical skill, but for the sake of the ingenious contriver himself, whose unwearied perseverance and high talents richly deserve it. With a rank and fortune, and every circumstance that usually unfit men for scientific pursuits, especially for their practical details, if he only encouraged those undertakings in others he would merit our praise; but when we see him, without losing sight of the duties of his station in society, give up so much time and expend so much money on those pursuits himself, and render not only his name illustrious but his rank more honorable, we must feel sympathy in his successes, and be rejoiced that he has

obtained from all quarters the very highest and most flat-
tering encomiums, and that he can now enjoy in the use
of his Telescope, the well earned fruits of all his previous
labours.

THE END.

REPORT

MADE AT

THE ANNUAL VISITATION

OF THE

ARMAGH OBSERVATORY,

BY REV. T. R. ROBINSON, D.D.

PRINTED

By Order of the Governors.

~~~~~~~~~~~~~~~~~~~~~~~~~~~

ARMAGH:

PRINTED BY JOHN M'WATTERS, ENGLISH-STREET.

1842.

REPORT, &c.

In laying before the Governors of the Observatory
a brief report of its present condition and the work
which has been performed in it up to this period, it
is my object to enable them to judge more correctly
than can be done in the ordinary course of a visita-
tion, how far this institution is fulfilling the intentions of its munificent patrons.

Astronomy, unlike most other branches of scientific research, cannot be effectually prosecuted, without appliances beyond the reach of private individuals: the instruments necessary for its investigations
are costly, and require extensive localities for their
erection; and the effectual employment of them
exacts an incessant devotion of labour and time,
which is incompatible with any other profession,
or indeed pursuit. Yet (such is its confessed supre-
macy in the world of intellect) even from the earliest ages, its wants have been supplied, and the calls
of the Astronomer for the means of Observation, have
been answered more promptly than those means
have been put in operation when obtained. The
primary business of an *Observatory* is obviously to
furnish *Observations;* yet, self-evident as this pro-
position is, it is frequently forgotten, and we may
say, with truth, that not more than a fifth of those

now existing, are fulfilling the real objects of their establishment. The causes of this are various : some of their astronomers may not have possessed the necessary combination of practice and theory, which is not very easily found ; others have, after a time, been wearied by the labour and uniformity of the pursuit; and more have turned away from their legitimate object to one which, though of higher order, ought not to interfere with it. In such Observatories as Greenwich or Pulkova, where the most extreme demands of the Director are conceded without limit, he possesses the power of advancing theoretical as well as practical Astronomy; but such cases are the exceptions, not the rule. In those of ordinary occurrence, it will be found that the two departments are incompatible; and there can be no doubt which ought to be the chief object. The theory of Astronomy can be pursued any where, or at any time—the facts on which it is built, can be obtained *in the Observatory alone.* Every hour that passes changes the face of the heavens, and if unobserved, is irrevocably lost. The Geometry of the present day will assuredly be eclipsed by that of our successors, and our theories be superseded by their more perfect developements; but good Observations increase in value with the lapse of time; and whoever holds such an office as mine, without affording them, will assuredly be branded by posterity as a *defrauder of science.* Impressed with this conviction, I have ventured to lay before the Governors the present statement, and to express my hope that they will *require* annually a similar one from me and my successors.

In respect of Instruments, the Observatory is now rich to a degree seldom equalled. It is, however, only necessary to mention the changes and additions which have been made lately*. The transit instrument, after 14 years' constant use, has its pivots as smooth as when they came from the maker. In 1839, they were very carefully examined as to their figure and equality. In the latter respect no difference could be found, and the extreme change, at different altitudes, is but 0".3. The level has had its tube replaced by one of double sensibility, by Ertel of Munich, and the surfaces of its contact with the pivots faced with agate, by Sharp of Dublin. In Feb. 1838, it was found that one of the quartz planes of the western Y was crushed at its extremity. A similar accident had happened to the circle, when Mr. Jones was redividing it. I replaced it and its fellow by slips of *hard* polished steel, and I find their action as good as that of the stone had been, so that I would recommend this material in preference. Nine wires are habitually used.

In 1832, I had applied to the transit clock the barometer tubes which I had contrived for the purpose of compensating the changes of rate caused by the airs varying pressure.† I found it impossible to obtain certain results, chiefly in consequence of irregularities in the arc of vibration; and, after

---

* The Transit is described in the introduction to the Armagh Observations for 1828 and 1829. The Circle, in the 9th Vol. of the Memoirs of the Royal Ast. Society, the West Equatorial is described by Troughton himself, in Rees' Cyclopedia under that heading. The East Equatorial is a Reflector of 15 inches aperture, used either as Cassegrainian or Newtonian, by Mr. Grubb of Dublin; its mounting was constructed as an experimental model of Mr. Cooper's great Equatorial, but only calculated to bear a ten feet Newtonian of Sir W. Herschel; it is, however, abundantly strong for its present load. The Circles being only intended as Finders, are but of 9 and 10 inches diameter, but it is capable of giving absolute places to a few seconds. It has a very effective clock movement and micrometer; and its optical power is such, that I have seen the 6th star of the Trapezium in Orion's Nebula. Each of these instruments has a good Clock belonging to it; the Observatory also possesses a 42 inch Achromatic, by Dollond, and a fine Sextant, by Troughten.

† Trans. Royal Ast. Soc., vol. 5.

three years, removed them, supposing the cause to be some unsteadiness in their attachment. The irregularity continued, however, and in 1840, was found to arise from the sapphire that communicates the action of the crutch to the pendulum, occasionally touching the bottom of the aperture in which it works. It was remedied at once, and the tubes replaced; their action is now steady, and has the remarkable effect of maintaining a uniform arc of vibration, and I am approximating to their true position. The circle has had eight additional microscopes applied to it by Mr. Dolland, for the purpose of correcting any errors of division that may be found in the mean of the original four. They might be all used in cases where extreme precision is required, but at a sacrifice of time far more than commensurate to the gain of accuracy. I have explained the method in my Memoir on Refraction (Transactions R.I. Academy, vol. xx.,) and will merely add, that the suspicion expressed there of the correction varying with temperature, has been in some instances strikingly confirmed. I have been obliged to remove one of the telescope micrometers, as it could not be used without disturbing the zero of the other. Gas light illumination has been introduced into the Observatory, to our great convenience; we have, however, found it necessary to make special provision for carrying off the products of combustion and the heat, which is evolved in greater quantity than by common lamps.

The dome covering the east equatorial ran on three bronze balls, between rings of oak, and though it moved well at first, yet as the wood was crippled by the pressure, it became so stiff that a force of 170 pounds were required to move it. A ring of cast iron, in 8 segments, has been fixed to the dome, and it runs on 12 bronze rollers attached to the wall-plate. This arrangement acts well, as a force of 30lb. moves it, and it can be commanded by an observer at the telescope, with a gun-tackle purchase. I may add, that I have found Lord Rosse's mode of polishing applicable to giving the hyperbolic figure to the small Cassegrain speculum, and that I am constructing an apparatus to polish a duplicate of the large one.

The west equatorial is nearly useless, from a similar difficulty, depending, however, on its original construction, As soon as the funds *admit* it, this defect should be remedied; at present, however, in consequence of the expenses which the transfer of the Kew instruments have made necessary, this cannot be effected.

These, which belonged to the private Observatory of Geoege III., who delighted in this science, and was himself an excellent observer, have been presented to us by Her Majesty. Some of them are merely valuable as remembrances of the past, others are of much greater value. Of these are two journeyman clocks, one a half-second time piece, which were much wanted; four other clocks, two of which, by Graham and Vulliamy, are to measure small fractions of the second. The best of them, by Shelton, I intend to replace our present mean time clock, but it has required considerable alterations, and a new scapement. Sir James South has also determined for me its barometrical correction, by placing it in vacuo. There are several telescopes, some very bad, some good. A complete one by Short, 2 feet focus, 6 inches aperture, with Newtonian, Gregorian and Cassegrain mirrors; a 10 feet,

8

9 inches aperture, by Sir W. Herschell, finished
with more care than usual, and defining well. An
achromatic of 8 feet, and $2\frac{3}{4}$ inches aperture, which
I have applied instead of a finder to the east equatorial, as the observations for determining the errors
of that instrument cannot well be made with the
reflector itself, since the mirror shifts in its box on
reversing the instrument. A transit instrument, $8\frac{1}{2}$
feet, and $3\frac{3}{4}$ aperture, which I hoped might super-
sede our actual transit; it however is of the same
optical power, and beyond comparison weaker in
its structure. There are, however, many conjunc-
tures in which a second instrument of the kind may
be useful, and for such it is reserved. The zenith
sector is, however, the most important part of this
acquisition. It is of the finest workmanship and
in perfect preservation. The object glass, of 12 feet
focus and $4\frac{1}{4}$ inches aperture, is excellent, defining
12 Lyncis very sharply with 300. The power originally belonging to it was 50, but I have had eye-
pieces and a micrometer constructed for it, by Mr.
Andrew Ross, the latter of which I consider a pattern
of excellent workmanship. A tower has been built
for its reception, over the eastern part of the transit
room, whose roof is in part moveable, and thus gives
a convenient place for occasional observations with
portable telescopes. The sector is carried by an
insulated stand of cast-iron, constructed by Messrs.
Gardner of this town; and effective means for illuminating the field and divisions applied. Indeed,
those originally belonging to it were so imperfect,
that I can scarcely believe it ever was used at night.
I have only recently obtained wire fine enough for
its plumb line, and have ascertained that its observations are capable of great exactness, but that the

combination of wood and metal in its construction is a defect which must be remedied. This, however, can easily be effected.

My report of the work performed with this magnificent collection, is, I regret to say, not quite so satisfactory as it would have been, had I more assistants. For more than half the time I have been labouring alone. My late assistant, from his infirmities, was able to give me little help, after the erection of the new transit : the difficulties attending the collection of the Observatory's income, in the following years, prevented me from pressing the appointment of a successor to him, till 1837; and since that, occasional illness and inevitable absence on both our parts, have produced many interruptions. Indeed, in respect of illness alone, there should always, if possible, be a second assistant. Still I have no reason to be ashamed of my doings. The line of Observation which I proposed, when the Primate gave the instruments which made it possible, was to redetermine the stars of Bradley's Catalogue in particular; and besides, of such others as, from their reputed accuracy and age, were likely to be of use, in reference to the proper motion of those bodies. I knew that the standard stars and the planets were carefully attended to at Greenwich, but this less attractive department was comparatively overlooked. To it, therefore, I applied myself as assiduously as the circumstances alluded to would permit. With the transit, up to the present time, we have recorded 20,000 observations and upwards, and with the circle, since its completion in 1834, more than 10,000, besides the number that were wasted, in consequence of its defects of division. I reckon that five observations give sufficient exactness, and therefore consider that by these 1600 stars have been finished in right ascension, and 1500 partially observed. A standard catalogue of 41 stars has also been formed, by direct reference to the equinox. The moon has also been frequently observed, and other planets occasionally. In polar distance 750 are completed, and 400 in progress, very few of which have (I believe) been observed in Britain. Many, also, have been observed below the pole, to form my tables of refraction, and 230 Observations of the sun have served to find the equinox and the obliquity of the eccliptic.

The mere observing is, however, the least part of the labour, for the observation itself is comparatively of little value unless reduced; the mere making it is, in fact, a pleasure, but the process which is necessary to make it a part of the currency of science is so irksome, that it scarcely ever finds any *volunteers* to engage in it. This may be illustrated by the statement, that supposing the constants of reduction known, the computations which must be performed before a transit observation is available to aid in forming a catalogue, require at least 250 cyphers; and that in most cases we have been obliged to compute those constants also, which require 350 more, and to do which, for 20 stars, is a good day's work. Much delay arose in this department, from the necessity of determining the constant of refraction for this Observatory, as a change in it would have required the revision of all. That was not effected till last year, and when I state that the circle reductions are finished to the end of 1840, and that I hope, in another year, to overtake the current observations, it will, I hope, be admitted that I have not been idle. The transit reductions are considerably more advanced, so that I think we are as little in arrear as most Observatories.

But the most important consideration remains, how these results, which have cost so much labour, and which I hope I may, without the imputation of vanity, think of some value, can be communicated to the world, " Paulum sepulto distat inertiæ celata virtus."

The easiest course is to publish the final deductions in the transactions of the Astronomical or some other Society. Astronomers, however, are never satisfied, unless they have the detailed Observations on which such deductions are founded; and with good reason. The elements of reduction, aberration, nutation, precession proper motion, are not absolutely determined : whatever future changes they receive can be applied as corrections; *effectually*, when the Observations are given in detail; but when only the result is given, doubtfully and imperfectly. This might, however, be remedied, though only in part, by depositing in the archives of the Society which publishes them, a manuscript copy of the originals. On the other hand, to publish them, I do not say as splendidly as the Greenwich or Cambridge Observations, but even as the three first years of mine were printed, is utterly beyond the resources of the Observatory fund, and ought not to depend on *private* liberality. I have, however, little doubt, from the encouragement which the Government now gives to science, that an application to it for the means of printing them, would be promptly attended to. It is, in fact, a national object, and one of the most honourable kind; and as the Observations of Edinburgh and the Cape are already printed at the public expense, the difficulty of precedent would not be in the way. Indeed, I think even more might be ventured, and if it were stated that this Observatory, established, and endowed in almost Royal magnificence, which has cost the Nation nothing, has been deprived of a fourth of its income by the Tithe Commutation, I am disposed to think we might even get the means of attaching to it a second Asssistant, for whom, independent of the contingencies alluded to, there would always be ample employment.

T. R. ROBINSON.

Sept. 15, 1842.

COMMUNICATIONS

BY THE

REV. T. D. ROBINSON, D. D., M. R. I. A., &c.,

ON

THE NEBULA, H. Fig. 44;

AND ON

AN ANCIENT BRONZE VESSEL FOUND IN THE KING'S COUNTY.

FROM THE PROCEEDINGS OF THE ROYAL IRISH ACADEMY, VOL IV. p. 236.

THE Rev. Dr. Robinson proceeded to notice a fact of some interest which he lately observed with the Rosse telescope. It related to a remarkable planetary nebula, Herschel's figure 44. This looks, in smaller instruments, like an oval disc, reminding one of the planet Jupiter; but it appears to be a combination of the two systems which he had formerly described. In both these the centre consists of a cluster of tolerably large stars: in the first, surrounded by a vast globe of much smaller ones; in the other by a flat disc of very small stars, which, when seen edgeways, has the appearance of a ray. Now this nebula, which he had recently observed through Lord Rosse's telescope, has the central cluster, the narrow ray, and the surrounding globe. He would also add, as a remarkable proof of the defining power of this vast instrument, that he saw with it, for the first time, the blue companion of the well-known γ Andromedæ, distinctly, as two neatly sepa-

A

rated stars, under a power of 828. It was discovered by the celebrated Struve, with the great Pulkova Refractor, and is a very severe test. He further wished to mention that, as La Place had anticipated, the ring of Saturn, which was quite visible, showed irregularities, which are most probably mountains, on its eastern side.

The Rev. Dr. Robinson then read the following communication, descriptive of the contents of an ancient bronze vessel found in the King's County, and now belonging to the collection of the Earl of Rosse. The antiquarian relics contained in this vessel comprised several celts, some spear-heads, gouges, and curiously constructed bells ; they were composed of a beautiful hard bronze, in very fine preservation. The composition of the metal itself, and the style of workmanship evinced in the various articles, argued no mean degree of metallurgic skill in their fabrication. Several of those interesting relics were exhibited to the members ; and drawings, which were pronounced to be admirable in their fidelity and minuteness, were displayed of the several implements of war and husbandry which were not exhibited.

" Several years ago, I remarked this vessel in the collection of the Earl of Rosse, and the singularity of its contents made me suppose a description of it might interest the Academy. I, however, found it impossible to acquire any information as to the locality or time of its discovery till now. It had been purchased for Lord Rosse, about sixteen years since, by an inhabitant of Parsonstown ; but the men who had found it, with that strange suspiciousness that is such a peculiar feature of the Irish peasant, had made him promise to keep the details secret during their lives. The last of them died this winter, and then Mr. —— felt himself at liberty to give me this information. It was found in the townland marked Doorosheath in sheet 30 of the Ordnance map of King's County, near Whigsborough, the residence of Mr. Drought, in what appears from the description to have been a piece of cut-out bog, about eighteen inches below the surface. No river is near the spot ; no bones or ornaments, or implements of any kind, were near it : though, had any gold or silver been discovered, the finders would probably not have acknowledged it to any one. I could not learn in what position it was found.

" A very good idea of the appearance of this vessel is given by fig. 1 of the accompanying drawing, for which I am indebted to Arthur E. Knox, Esq.* The scale is one-third of the original, and he has given very precisely the actual condition of its surface. It is composed of two pieces, neatly connected by rivets. The bronze of which the sheets are formed possesses considerable flexibility, but is harder than our ordinary brass ; and it must have required high metallurgic skill to make them so thin and uniform. On the other hand, it is singular that, neither in this nor any other bronze implements with which I am acquainted, are there any traces of the art of soldering : if it might be supposed objectionable in vessels exposed to heat, yet in musical instruments this would not apply.

" Such vessels have often been found, but the contents of this are peculiar. When discovered (without any cover) it seemed full of marl, on removing which it was found to contain an assortment of the instruments which may be supposed most in request among the rude inhabitants of such a country as Ireland must have been at that early epoch. A few were given away, one of each, in particular, to the late Dean of St. Patrick's, and these are probably in our Museum ; but the following remain :

" 1. Three hunting horns, with lateral embouchure, shown on the scale of one-third at fig. 2 (D. 656).

---

* This vessel is very similar to one in the Museum of the Academy, which is marked D. 551. As several of the other objects described by Dr. Robinson resemble the specimens contained in the Museum, a reference to the latter is given in each case.

" 2. Ten others of a different kind, fig. 3 (D. 653) : these differ considerably in size, but that represented is of the average size. Some of the largest have the seam united by rivets; in others it is marked by a paler line in the bronze, which seems as if they had been brazed, but is probably owing to a thin web of metal, which penetrated between the halves of the mould in which they were cast. All of this kind seem to have had additional joints, of which three were found, figs. 4 and 5 (B. 963) ; at least, no other use of these pieces occurs to me ; and in none of them is there any convenient embouchure.

" 4. Thirty-one bells of various sizes, figs. 6 (B. 945) and 7 being the extremes; of the real size. They have loose clappers within, and many of them slits to let the sound escape more freely. The bronze in these is much harder than in the preceding, and has resisted decomposition almost entirely. I think it can scarcely be doubted that these were bells for cows and sheep, which would be specially useful among the dense forests which then overspread the island.

" 5. Thirty-one celts, of very different sizes, but none sufficiently large to induce a belief that they were used in war. In many of them the colour of the bronze is such as, at first sight, to excite an opinion that they were gilded. There are

Two of the size of fig. 8 (B. 244).
Seven of the size of fig. 9 (B. 347).
Six of the size of fig. 10 (B. 350).
Five of a size intermediate between these, and
Six of the size of fig. 11 (B. 270).
Five of the size of fig. 12 (B. 276).

" It is worthy of notice that in all the points are entire and sharp, and the edges unbroken, and not seeming to have been ever used.

" 6. Three gouges, fig. 13 (B. 181). These are, I believe, of comparatively rare occurrence, and therefore were, probably, of less extensive use than the celts : just as the common carpenter's gouge is with respeet to his chisel, to which I believe the others to have been the analogues. Their round edge is well adapted either for paring or for excavating bowls and goblets.

" But the finest specimens of workmanship are the spears, twenty-nine in number. These also are of various sizes, and of greater diversity of pattern than the other implements. There are

Two of the size of fig. 14 (B. 54).

Four of the size of fig. 15 (B. 38).

One of the size of fig. 16 (B. 35).

Seven of the same size, but a plainer pattern.

Nine of the size of fig. 17 (B. 34).

Six of a size two-thirds the preceding, but which it did not seem necessary to draw.*

---

* It is a curious circumstance that *six* kinds of spear-heads should have been found. Dr. R. had met with seven different names for this weapon in Irish; but as his knowledge of this language is very limited, he availed himself of the high authority of Mr. Eugene Curry, who gives them :

Laıʒhın,	pronounced Loy-en.
Sleaʒh, ...	Shle.
Manaır, ...	Mon-eesh.
Cnuıreach, . . .	Crusheach.
Poʒha, ...	Fow-gha.
Ʒae, ...	Gaé.
Ʒabal, ...	Ga-val.

With the remark, however, that Sleaʒh and Ʒae are sometimes used indiscriminately. The Laıʒhın was of foreign introduction, and peculiar to the men of Leinster; it was, therefore, not likely to be used in this locality, so that the collection, probably, comprised all that were in demand. Among these names, four are evidently of Hebrew affinity. The second is identical with שלח (shlech), a missile spear ; the third comes from מנה, fate ; the fourth, or rather its abbreviate form, Cnuıth, is from כרת (chreth), to destroy ; the sixth is little altered from קין (kain), a dart; and the last, possibly, comes from גוב, to divide. Mr. Curry remarks, also, that several of these names are now given to agricultural instruments ; the loy and slaine are familiar examples : Manaır now means a mason's trowel. It should seem that metallurgy was made the minister of war long before it became subservient to the arts of peace.

" These, also, have their points and edges perfect, and seem never to have been used; they show not only that the work-men who made them were perfect masters of the art of casting, but also that they possessed high mechanical perceptions. If these weapons and the bronze swords (of which our Museum contains several) be compared with those used in our army, it will easily be seen that the former are constructed on principles far more scientific. Some of these may not be obvious to the ordinary reader, as they depend on the properties of bronze. This alloy, especially when in the proportion used for weapons (in which it is an atomic compound, containing fourteen equivalents of copper and one of tin, or nearly eighty-eight and twelve by weight, and possesses a maximum specific gravity considerably surpassing either of its elements), combines great strength and toughness, but has not hardness to take an effective and permanent edge. It has, however, been shown by D'Arcet, that if its edge be hammered till it begins to crack, and then ground, it acquires a hardness not inferior to the common kinds of steel, and is equally fitted for cutting instruments. Now, in fig. 14, the strong central cone of bronze, remaining in its ordinary state, effectually stiffens the weapon against fracture; while the thin webs on each side have evidently been subjected to this or some similar process, for their edges are much harder, as well as brittle. In the smaller weapon, fig. 17, the web might be too thin, and, therefore, it is reinforced by a pair of secondary ribs; and in fig. 16, the most highly finished of them all, by four such. It is, however, possible that these ribs may have answered another purpose; they have so strong a resemblance to those on some Malay krises, that they may have been designed, as in those weapons, to retain poison. This practice, I fear, was not unknown among the ancient Irish, as, indeed, it seems to have prevailed among all the Celtic and Iberian races; thus, in the poem on the death of Oscar, published by Bishop Young, in the first volume of our Transactions, the spear of

Cairbre is expressly said to be poisoned (Nime), and nothing seems to require a figurative sense of this epithet to be understood.

"The most obvious hypothesis respecting this curious assemblage of objects is, that they were the property of some individual, who concealed them in the bog, perhaps on the approach of a predatory party, and perished without recovering them. Against this is the fact that the tools and spears seem not to have been ever used, and the probability that, in such times, every spear-head would have been mounted, and in the hand of a combatant. It seems more likely that the collection was the stock of a travelling merchant, who, like the pedlars of modern times, went from house to house, provided with the commodities most in request; and it is easily imagined that, if entangled in a bog with so heavy a load, a man must relinquish it. And this is connected with another question, the source from which the ancient world was supplied with the prodigious quantities of bronze arms and utensils which we know to have existed. This caught my imagination many years since, and I then analysed a great variety of bronzes, with such uniform results, that I supposed this identity of composition was evidence of their all coming from the same manufacturers. Afterwards I found that the peculiar properties of the atomic compound already referred to are sufficiently distinct to make any metallurgist, who was engaged in such a manufacture, select it.* It also appeared to me more permanent in the crucible than when of higher or lower standard. But the same conclusion is forced on us from another ground. Bronze contains tin; now this metal, for all commercial purposes, may be said to be confined to the

---

* The technical importance of atomic proportions is remarkable. Speculum metal is 4Cu + 1St; gong metal is 8Cu + 1St; that referred to is 14Cu + 1St; the hardest metal used for cannon is 16Cu + 1St.

south-west of England,* and, therefore, the bronze trade must have originated with persons who were in communication with Britain. But in ancient history we find only one people of whom this can reasonably be supposed, the Phœnicians, who, like ourselves, seem to have been the great manufacturers and merchants in olden time. That they had factories, if not colonies, in Spain, at a very early period, is known to all; and it seems most unlikely that such enterprising navigators would stop there. Of course, one can attach little weight to the remote traditions of Irish history, if unsupported by other probabilities; but the traces of Phœnician intercourse which they exhibit are borne out by the admixture of Punic words in the language, and by usages which show that the worship of the god Baal, and other Sidonian rites, had once prevailed in the island. Their traffic in amber proves that they must have gone yet further, even to the Baltic; for then, we may be sure, the land carriage of precious materials through various and hostile regions was almost impossible. All, too, that we know of early antiquity shows that they had the bronze trade in their hands. Even down to the time of Aristotle, tin was described by the epithet 'Tyrian;' and in every nation where bronze was in common use, their presence can be traced or inferred. In Egypt, where this compound was of universal use, we know that the people were little addicted to maritime pursuits; while they were in close communication with the Sidonians (of the same race), through the Mitzräite colony of the Philistines. In Etruria, not less remarkable for its profuse employment of bronze, we know that they did not obtain it directly, for it is recorded that an expedition was fitted out by them, to open a

---

† There are tin mines in Malacca, but we have no evidence that they were worked so early; and if they had been, it is quite improbable that their produce found its way to the Mediterranean.

communication with the tin islands, which failed, in conse-
sequence of the jealousy of the Phœnicians. Hence we may
conclude that the latter held a monopoly of the tin. In Judea,
we find Solomon obliged to employ a Tyrian founder for the
bronze works of the temple, and we gather from the account,
also, how they were cast—in loam.* But Greece, in the Ho-
meric age, presents a state of things much more conformable
to what I suppose was the condition of Ireland when this collection was buried. Iron scarcely appears to be in use; and
it may be surmised that the art of working bronze itself was
not generally understood, from the poet's description of Vul-
can making the arms of Achilles. No mention is made of
casting or moulds, though a reference to Milton's splendid
description of the infernal palace shows how much more
poetic that would have been than the hammer and anvil. It
seems as if the god merely heated and chased into shape
sheets of metal, already prepared.† It may be added, that
Homer describes all articles of superior workmanship as Sidonian; and represents this people as trading in every part of
Greece. Their ships run into some cove, and their factors
go to the dwellings of the neighbouring chiefs. These,
though at feud among themselves, and driving each other's
cattle on every opportunity, receive the strangers kindly, and
purchase from them hardware, jewellery, articles of dress, and
toys, in return for cattle and slaves. Now, just such a person
I suppose the possessor of this vessel to have been, and of this
very nation. Commerce was probably carried on in this
way along the shores of the Mediterranean, till the destruc-

---

* Moulds for celts have been found here and in other countries, but were,
perhaps, employed to recast old bronze; they could not turn out work very
neat, and many of these tools have apparently been cast in sand. These spears
were, I think, cast in loam.

† Bronze is brittle at a red heat; but it and even bell-metal are mallea-
ble at a temperature below visible ignition. Speculum metal is not brittle
while red hot.

tion of Tyre by Nebuchadnezzar destroyed it also for a time, and then removed its most powerful centre of action to Carthage. That state seems to have chiefly directed its attention westward; and it is a confirmation of my opinion that the bronze trade was almost exclusively Phœnician, that about this time the use of the alloy rapidly gave way to iron and steel. In fact, the supply being cut off from Greece and Asia by the ruin of the Tyrians, *they* were obliged to seek other resources; but in Ireland and other Atlantic lands the traffic must have continued, nay, perhaps, even increased, in consequence of that event, till the fall of Carthage finally cut it off. I would also throw out another suggestion, though at considerable risk of being thought a dreamer. We see in Homer that the Phœnician traders were quite ready to have recourse to violence when they could profit by it; and, from more historic sources, that, in Lybia and Spain, they took an early opportunity of turning their factories into forts, and enslaving the natives. Did the same thing happen *here*, when the Tuatha De Danaan, a tribe rich in metallic ornaments and weapons, subdued the ruder Firbolgs, who referred their superior knowledge to magic? Were these shadowy personages also Phœnicians? Their name signifies " the tribe of the gods of the Dani or Damni." If the first, it might indicate Odin and his Asæ; but, besides that *they* must have been far later, it seems highly improbable that such fierce warriors would have been overpowered by any Celtic immigration. If the second, the Damni, the inhabitants of Devonshire and Cornwall, must have been completely under the influence of the Phœnician agents, and may at first have imagined and called their accomplished visitors deities. In these Ogygian regions we must not reckon dates too closely; but I believe it is held that the battle of Moytura, which established their dominion, was fought about 600 years before our Lord, and, therefore, at the very time when the fall of Tyre may have been supposed to scatter its people, and the ruin of their commerce incline them to desperate adventure. It is possible that this conjecture may be established or disproved by a comparison of the skulls found in the sepulchral monuments on their battle-fields with those of Tyrian or Carthaginian origin, if any such are known to exist."

THE END.

A

DESCRIPTIVE AND EXPLANATORY

ACCOUNT OF

THE EARL OF ROSSE'S

REFLECTING TELESCOPES:

OF THE

PROCESSES EMPLOYED IN THEIR CONSTRUCTION;

AND OF OBSERVATIONS OF SOME OF THE

NEBULÆ

MADE WITH THESE INSTRUMENTS.

———————

ILLUSTRATED BY ENGRAVINGS.

———————

LONDON:
THOMAS RICHARDSON AND SON,
172, FLEET STREET, 16, DAWSON STREET, DUBLIN, AND DERBY.

PREFACE.

THE following sketch has already appeared in the *Dublin Review*. The writer accedes to a wish expressed for its republication, more on account of the great importance of the subject, than from a hope that he has done it anything like adequate justice. The interest which attaches to the labours of the Earl of Rosse can no longer be regarded as of a private and personal character: it belongs to the history of Astronomical Science in Ireland; and perhaps, in the absence of a more complete treatise, a concise and popular narrative of his experiments in the improvement of the astronomical telescope, accompanied by a comparative summary of the similar efforts made before his time, may lead to a more correct appreciation of the eminent success which he has attained, and which reflects the highest credit, not only upon himself, but upon the whole Irish scientific community.

The present short essay does not pretend to anything more than this. Seeking to popularize the subject, as far as is consistent with its very scientific character, the writer has thought it better to abstain from those technicalities which might embarrass the general reader, and from that closeness and precision of phraseology, which, though indispensable in a strictly scientific treatise, and perhaps necessary for full and perfect accuracy, would be out of place in a purely popular dissertation.

The illustrations are taken, partly from a very excellent pamphlet printed at Parsons-town, partly from a paper by the Earl of Rosse in the *Philosophical Transactions* for December, 1844.

THE EARL OF ROSSE'S

REFLECTING TELESCOPES. *

IN the Museum of Natural History at Florence, is preserved a relic, which, from its mean appearance and
diminutive size, may possibly escape the notice of a casual
visitor. † And yet we doubt whether Florence, among all
her other priceless treasures of ancient and modern art,
possesses any thing half so interesting to the lover of
science. It is the TELESCOPE OF GALILEO—the very
instrument with which, as the inscription records, "he
discovered the spots of the sun, the lunar mountains, the
satellites of Jupiter, and, as it were, an entire new uni-

---

* The facts relied on in the following pages are derived from the following
sources:

The Monster Telescopes, erected by the Earl of Rosse, Parsons-town, with an
Account of the Manufacture of the Specula, and full Descriptions of all the
Machinery connected with these Instruments. Illustrated by Engravings.
Second Edition, 8vo. Parsons-town: 1844.

Address of the Earl of Rosse to the British Association at York, on Monday,
September 30, 1844, on the Reflecting Telescopes lately constructed by him.
(Athenæum, Oct. 5.) London: 1844.

Observations on some of the Nebulæ. By the Earl of Rosse, F.R.S., &c.
(Reprinted from the Philosophical Transactions of the Royal Society of London
for 1844.) 4to. London: 1844.

An Account of Experiments on the Reflecting Telescope. By the Right
Honourable Lord Oxmantown, F.R.S., &c. (Philosophical Transactions for 1840.)
London: 1840.

† We believe this is hardly possible since the new and tasteful arrangement of
the gallery. We speak of what was the case some years back.—The inscription
is by Lanzi, and is worth transcribing:
"Tubum opticum vides, Galileii inventum et opus, quo maculas solis, et exti-
mas Lunæ montes, et Jovis Satellites, et novam quasi rerum universitatem,
primus despexit. A. MDCIX."

verse;" casting new light on the most mysterious phenomena of astronomy, substituting observation for conjecture, and banishing for ever the visionary theories and
pedantic forms, by which truth had hitherto been dis-
figured and concealed. And though the deductions of
the illustrious inventor himself were, in some respects,
halting and imperfect, it is true, nevertheless, to say,
that to him we are ultimately indebted for all that has
since been discovered. For, to use the eloquent illustration of one whose very name is an authority on such a subject, "this tiny combination—a telescope which the observer held between his fingers, or hid in the hollow of his
hand—was the mustard-seed of those mighty trunks which
now rise majestically to the heavens, and on which the
astronomer perches himself, like the eagle upon the lofty
cedar, to obtain a nearer glance of 'the God of Day!'"

It requires no ordinary effort of the mind to pass from
the telescope of Galileo to the telescopes of modern times,
and to follow the successive links in the chain of improvement by which these extremes are connected. What an
infinity of industry and genius—how many days of toilsome
study, how many nights of watching and anxiety, fill up the
interval! How would the fiery Florentine, as he gazed in
fond triumph upon his little tube, have stared in incredu-
lous amazement, had he been told of the "monster telescopes" of our day—so unwieldy as to be moved only by
the most massive machinery, and so enormous, that one of
the dignitaries of the church—seldom very shadowy personages—could *walk with outspread umbrella through the
tube!* * And yet, prodigious as is this contrast of dimensions, the contrast of results is still more wonderful.
Galileo thought it an extraordinary triumph—and so it
undoubtedly was—to discover the lunar mountains.—Lord
Rosse's three-feet reflector is reported capable of showing
any object on the moon's surface of the size of one of our
ordinary public buildings!† The range of Galileo's telescope was practically confined to the nearer bodies of our
system; its effect on the more distant, was hardly perceptible.—In the modern instruments, the observer scarcely
dares to expose his eyes to the blaze of the nearer lumi-

---

* "The Dean of Ely walked through the tube (of Lord Rosse's six-feet
reflector) with his umbrella up!" *North British Review*, No. iii. p. 207.

† Dr. Robinson's Address to the British Association at Cork, 1843. *Athenæum*,
No. 830, p. 867.

naries!* And thus, while Galileo's telescope is now almost exclusively confined to the nearest objects, and degraded to the humble uses of a mere opera-glass, its more ambitious successors walk abroad

"Throughout those infinite orbs of mingling light;"

prying into the most distant realms of space—regions of which Galileo, all daring as he was, never dreamed, even in his wildest speculations—distances for which the nomenclature of arithmetic can hardly find an expression, and of which, even when represented to the eye by long lines of figures, the mind can form but a vague and indefinite conception!

We should ill discharge our duty of chronicling the progress of science and literature in Ireland, did we omit to record an event which is destined to form an epoch in the history of astronomy—the successful termination of the experiments on the improvement of the reflecting telescope, in which our distinguished countryman, the Earl of Rosse, has been engaged for a series of years. We had intended to defer our notice till the performance of the gigantic instruments which he has constructed should have been fully tested. But in deference to the anxiety universally expressed upon the subject, we have been induced to abandon our first intention; and we, therefore, proceed, at once, to lay before our readers, as far as shall be compatible with the limited space at our disposal, the most important results which have been hitherto obtained. The subject is one which it is difficult to divest of technicality; we shall endeavour to avoid it, as far as this may be possible, confining ourselves to a plain and popular account of the processes employed in the construction of the telescopes, the principles by which they have been guided, the difficulties which have been successfully surmounted, and the principal facts elicited by the very limited trial which, as yet, it has been found possible to give the instruments since the time of their completion.

To those who have studied the progress of any of the

---

* "Intent on far discovery, Herschel seldom looked at the larger stars; and, because their blaze injured his eyes, he rather evaded their transit. But at one time the appearance of Sirius announced itself at a great distance like the dawn of morning, and came on by degrees, increasing in brightness, till this brilliant star at last entered the field of the telescope with all the splendour of the rising sun, and forced me to take my eyes from the beautiful sight." *Nichol's Architecture of the Heavens*, p. 31. The same is even more true of Lord Rosse's three-feet telescope.

modern arts—who have watched the prodigious advances
of the steam-engine, the rapid development of practical
chemistry, the ingenious refinement of almost every mechanical process, the power-loom, the paper-machine, the
printing-press—it may seem strange, that, after a period of
nearly two centuries and a half, during which it has un-
ceasingly occupied the most acute minds and the most skil-
ful hands in every country of Europe, the perfection of the
telescope should still remain an unsolved, perhaps, an in-
soluble problem. To understand this seeming anomaly,
it might be enough to remember, that the objects upon
which the other arts are employed, are immediately under
the hands of the artist, and, at least indirectly, control-
lable at his pleasure; while the field of telescopic action is
amid the distant spheres, far removed beyond the reach,
not alone of the hands, but almost even of the imagination, of the artist. But no curious inquirer will rest here.
In order to appreciate fully, and, indeed, even to comprehend, the nature of the operations on which the improvement of the telescope depends, and to which Lord Rosse's
eminent success is attributable, it will be necessary to
review, briefly, the difficulties which lie in the way of the
perfection of this important instrument; and to describe
the various efforts by which, at different times, it has been
sought to overcome them. If, in thus accommodating our
observations to the convenience of the general reader, we
shall be compelled to dwell upon things which must appear
almost as first principles to our scientific friends, we can
only hope that the motive may be accepted as our apology.
It might seem at first sight, that, the principle of the
telescope once discovered and applied, the perfection of
the instrument was comparatively easy. If the combination of a convex and concave lens (Galileo's telescope), of a
given magnifying power, had the wondrous effect of causing dim and distant objects to appear more brilliant and
nearer to the observer, it would not seem unreasonable to
expect, that, by increasing the magnifying power of the
lenses, the effectiveness of the telescope might be indefinitely increased; and as it was known, that the magnifying
power of the lens increases in certain proportions with its
aperture or surface, the first sanguine conclusion was, that,
were it only possible to procure lenses of considerable size,
the whole work was accomplished. And, indeed, the theory
of vision with a telescope would seem to countenance and

confirm this expectation. Though the explanation is far
from complete, our meaning will be sufficiently understood
when we state, that the eye is said to see distant objects,
whenever the rays of light which emanate from them are
sufficiently numerous and powerful to affect the retina in
that unexplained and mysterious way which is called the
sensation of sight. Now, as the rays emanate from a luminous body in straight lines, which diverge from each other
in every direction, it is plain that the divergency of the
rays from each other, or the thinness of the light, increases
in proportion to the distance of the body; and, consequently, that the number of rays falling within any given
space proportionally diminishes. As the number of rays
falling within the aperture of the pupil diminishes, the
retina is less sensibly affected, and the apparent brilliancy
of the body declines; and when the number ceases to be
sufficient to affect the retina *at all*, the body becomes practically invisible to the observer. This invisibility, however,
is clearly *relative*. The object, in such circumstances, is
invisible to the observer, not because *no* rays from it reach
the place where he stands, but because, in consequence of
their extreme dispersion, the number which falls within the
pupil of his eye is practically insufficient to affect the retina
with the sensation of seeing. And if the aperture were
increased, the number of rays received within it would be
increased, and the sensation of vision would be at once re-
stored. And such is the very provision by which nature
produces this effect. By a spontaneous action of the
delicate mechanism of the eye, as the quantity of light
diminishes, the aperture is enlarged; and every one has
observed, that the night birds, as the owl and bat, which
voyage in the dark, are provided with enormously dilated
pupils, through which a greater number of rays are introduced into the eye, and, being collected into a focus, are
made to act upon the retina. Now, if a medium of such
size as to collect a sufficient quantity of rays, and possessing the power of causing them to fall in this collected form
upon the eye, be placed in front of the pupil, the effect will
be the same,* as though the pupil had been proportionately enlarged. And thus the object-glass of the refracting

---

* This is not strictly true, as light is lost both in refraction with a lens, and
reflexion with a mirror. In the former, the light received is, to the light transmitted after refraction, :: 1 : .938. In the latter, the light received is, to the light
transmitted after two reflexions, (which occur in all reflecting telescopes except
the Herschelian) :: 1 : .45242.

telescope, or the speculum of the reflector, is, to speak
loosely, nothing more than an artificially enlarged pupil;
and the quantity of light thus conveyed to the eye is as
much greater than that received without their aid, as their
aperture exceeds that of the natural pupil for which they are
substituted. Thus, too, it is, that the distance of objects
is apparently diminished by the use of such lenses. For
we judge of distance by the sensation. Now the sensation
produced by the artificially increased quantity of light in-
pinging upon the retina, is precisely the same as what we
should have, were we brought so much nearer to the object
as to receive, naturally, the same quantity of light within
the unassisted aperture of the pupil.

Hence, a hasty application of the analogy from this
natural process might lead one to imagine, that, as the
only apparent provision which Nature makes for increased
vision is the enlargement of the pupil, so, in order to produce an indefinite increase of magnifying power with the
telescope, it would only be necessary to procure lenses of sufficiently large diameter. But, unfortunately, the materials
at our command are very different from those which Nature
employs in her seemingly simple, but most mysterious
operations. Unfortunately, too, our knowledge even of
those materials which we possess is limited and conjectural;
and, even such as it is, is only to be acquired by long and
laborious research, embarrassed by many an unexpected obstacle, and embittered by many a harassing failure. To the
perfection of vision with the telescope, certain conditions
are clearly necessary;—it must show the object well
magnified, brilliant, and, above all, distinct and in its
true shape; and although Nature might seem, at the
first blush, to make no provision for the attainment of
these conditions, beyond the enlargement of the pupil,
yet a nearer scrutiny of her process detects a provision
for securing them all, so exquisitely delicate, and so perfect in its entire organization, as almost to baffle the
hope of imitation. In order, therefore, to construct a good
telescope, the artist must provide for all these conditions.
If he had to deal with them singly, his task would be
comparatively easy; but in extending the efficiency of the
telescope, he must preserve the collective proportions of
them all. He cannot purchase magnitude in the image at
the expense of true figure or of brilliancy; nor, above all,
can he sacrifice distinctness to any other condition, how-

ever important for its own sake. Now it unfortunately happens, that the means by which some of these properties would be secured, are precisely those which tend to diminish, and ultimately to destroy, others equally indispensable. Thus it is plain, for example, that with a given quantity of light, any attempt to *magnify* the image must diminish its *brilliancy;* for if the rays be dispersed over a greater space, the number in each particular point will be proportionately less.

This, however, is a difficulty which science would soon overcome, had she unlimited command of the requisite materials. But, unluckily, the artist is met at his very first step towards increasing the magnitude and brilliancy of the image, by enlarging the aperture, by other difficulties which arise from the nature of the media which he is constrained to employ. The learned reader will anticipate, that we refer to those embarrassing phenomena known among opticians by the names of *spherical aberration* and *chromatic aberration.* It would carry us quite beyond our limits to enter into their nature ; we must content ourselves with briefly indicating their source.

In order that a true image should be produced by a refracting lens or a mirror, it would be necessary that all the rays should be accurately collected into one focus after refraction or reflexion. There are certain geometrical figures (the paraboloid and the ellipsoid), by which (if it were possible to attain them with perfect mathematical precision) accuracy of reflexion would be secured. But it is found so difficult in practice to impart these figures to lenses and mirrors, that artists are generally compelled to content themselves with a good spherical form. Now, in a spherical lens (and the same is true of a speculum) the centre and adjacent parts form their image in a nearer focus than the circumference and the parts near the circumference. The consequence is a confusion of images, somewhat similar, except, of course, on an infinitely minute scale, to what any one may see when the shadows of the same object, formed by rays from different lights, meet upon a wall. This is called *spherical* aberration, because it is a consequence of the use of spherical lenses and mirrors.

The second—chromatic aberration—is still more serious. Every person is acquainted with the fact, that white, or compounded light, is not homogeneous, but may be resolved into various kinds, distinguished by Newton into seven, and by others into four, primary colours. Now, the rays of different colours are not all equally refrangible by the same medium; and the difference of their refrangibility by glass (the substance ordinarily employed for lenses), is the cause of a very material imperfection of the image, which is called (from the name) *chromatic* aberration. It is found, that the red rays are the least refrangible, and the violet, the most so; the refrangibility of the rest, being intermediate between that of the two extremes. From this unequal refrangibility it results, that the rays of the several colours cannot be all collected into the same focus; the most refrangible (violet) form their image nearest the lens; the least refrangible (red) form it most remote; the rest, at intermediate intervals. And the consequence is, that, though all have their foci in the axis of the lens, yet this axis, instead of presenting one unbroken and distinct image, shows a series of images, formed more or less perfectly by the separate rays, all blended confusedly together, and surrounded by a coloured fringe.

Such are, briefly, the sources of the confusion and indistinctness of images known under these names. They have been the great obstacle to the progressive improvement of the telescope; and it is to their mitigation or removal, that all the efforts of successive artists have been directed. The history is among the most interesting in the entire study of optics.

The cause of the first—spherical aberration—was early discovered. In Galileo's telescopes it was not very sensible, because as they consisted of a convex and concave lens, the opposite curvatures corrected each other's aberration. But as soon as the concave lens was dispensed with, it at once became apparent. It is inconsiderable, when compared with the chromatic aberration; still, as it increases rapidly with the aperture of the object-glass,* it was a serious obstacle to the enlargement of this—the great source of telescopic power. It decreases, however, as the focal length of the object-glass is increased,† and, hence, we find the first effort to correct this imperfection in the enormous focal length of the telescopes of the first epoch. Galileo had no occasion to employ this device;

---

* It varies as the square of the diameter.

† When the aperture is constant the aberration varies inversely as the focal length.

but the principle was soon applied by those who followed
in his steps, especially by Giuseppe Campani, Divini, and
Huygens. The first of these constructed lenses of 34, 86,
and even 140 feet focal length. Divini, too (celebrated
for his obstinate and dogged denial of Huygens's discovery
of Saturn's Ring) constructed telescopes of very long focus.
But both were far surpassed by Christian Huygens, whose
telescopes reached a focal length of 100, 120, 156, 174, and
even 210 feet; and still more by Auzout, who constructed
a glass (which, however, he never was able to mount) of
no less than 600 feet focus! To facilitate the manage-
ment of such enormous machines, Huygens adopted the
idea of the celebrated Nicholas Hartsoeker, a native of
Gouda, who separated the eye-glass from the object-
glass, placing the latter in a stand upon the roof of a house,
and thus dispensed altogether with a tube. Huygens improved upon the plan. He erected his object-glass upon the
top of a high mast, in a short tube moveable upon a ball
and socket by means of a cord, and thus easily brought
into a line with a small tube containing the eye-glass, which
he held in his hand.

Such were the first efforts to counteract the intractability
of the materials of the telescope. They reflect great credit
on their inventors; but they were necessarily very imperfect. Not to speak of the difficulty of managing an apparatus so enormous and so unsteady, the best result was
only a mitigation of the evil. Though the image thus obtained was free from any very sensible spherical aberration,
yet it was impossible to apply to it eye-glasses of such high
magnifying power as is necessary for minute astronomical
observation. The imperfections of the image, though partially insensible with low-powered glasses, became more
sensible as the power was increased; and, when glasses of
great capacity were employed, the indistinctness became
fatal to observation. We need hardly add, that this device
affected only the spherical aberration, and left the still more
serious defects arising from chromatic aberration entirely
untouched.

We pass, therefore, to the second epoch in our history.
The cause of the chromatic aberration was unknown till
the time of Newton. That he should at once divine it, was
a direct consequence of his immortal discovery of the composition of light. But even at the very moment when his
acute mind suggested, in the combination of water with a

glass lens, a remedy for the spherical aberration, very analogous to that since adopted for the chromatic, he hastily concluded that it was impossible to overcome the latter. He, therefore, pronounced the refracting telescope incapable of perfection ; and as the idea of reflecting astronomical telescopes had just been started by Gregory, he gave himself, with all his characteristic energy, to the work of their improvement. We shall so far depart, however, from the chronological order, as to continue the history of the refracting telescope to our own time, before we enter upon that of the reflector, to which class the telescopes of Lord Rosse belong. We shall thus be spared the necessity of a great deal of repetition.

When we said, at the commencement of this paper, that the materials at the disposal of our mechanists are very different from those which Nature employs in her own operations, we alluded to the natural construction of the eye by which both spherical and chromatic aberration are avoided. The provision for the correction of the latter is extremely simple and beautiful. The rays of light on their way to the retina, pass not through one single refracting medium, as in the telescopes hitherto described, but through several distinct media—the cornea, the aqueous humour, the chrystalline lens, and the vitreous humour. Now, these media possess different, and, in some respects, opposite refracting powers for the different coloured rays. Their refractions, therefore, mutually counteract each other, and entirely destroy the chromatic aberration.* This beautiful arrangement suggested the idea of forming a refracting lens, not as of old, of a single medium, but of two or more, the *dispersive powers*† of which (as in the human eye) might counteract each other. This ingenious

* We should add, that, by the natural construction of the eye, the spherical aberration is similarly obviated.

† The dispersive power of any refracting medium, is its capacity of separating the different coloured rays of a beam of light, and throwing them into distinct foci.

That we may not interrupt the history, we shall throw one or two necessary explanations into a note. The glass used by Newton *dispersed* a ray falling upon it at an angle of 30°, in such a ratio, that the extreme *red* ray emerged at an angle of 50° 21¼, while the extreme *violet* ray emerged at an angle of 51° 15¾. Now taking 0,5 as the sine of 30° (the angle of incidence), the sine of the emergence of the *red* ray will be 0,77, and the sine of the emergence of the *violet* ray will be 0,78. This will give the dispersive power of the glass which Newton employed. The dispersive powers of crown glass and flint glass (which are used to correct each other in achromatic lenses), are related to each other (nearly) :: 0,05 : 0,033. The focal lengths of the glasses, thus combined into one lens, must be in the proportion of their dispersive powers.

thought was first reduced to practice, in 1733, by an English country gentleman, Mr. Hall. But his invention never obtained much publicity, and his fame is now merged in that of the celebrated John Dollond, who was the first to construct, in 1758, achromatic lenses, of a considerable size, and for sale. By combining flint glass and crown glass in certain proportions in his lenses, as the dispersive power of the former is about one-half greater than that of the latter, this ingenious artist was enabled to neutralize the chromatic aberration, and to cause the different coloured rays in each pencil to unite in the same focus after refraction. The achromatic lens thus formed, however, was not entirely successful. Though the spectra formed by the opposite dispersions of the coloured rays in a lens made of flint and crown glass are equal to each other, they are not precisely similar. That is, though the *whole* coloured spectrum in each case is equal, the *coloured spaces* occupied by the *different rays* are not equal, and do not precisely coincide ; and the result is, that the image of a luminous object formed by such a lens, is found to be bordered on one side with a purple, and on the other with a violet fringe. This defect, however, has since been surmounted by Dr. Blair, who, by inclosing muriate of antimony between two convex lenses of crown glass, has produced a perfect compound lens, which refracts parallel rays to a single focus without a trace of this secondary colour created by the single combination of the glasses.

The difficulty of obtaining good glass, however, prevented the extensive adoption of these beautiful devices. The largest achromatic glasses in England at the beginning of this century hardly reached in aperture a diameter of six inches, and the enormously superior apertures attainable in the reflecting telescopes for a long time threw a damp on the attempts at competition. But at length, by one of those fortunate combinations of necessity and genius, to which science owes many of her brightest discoveries, a Swiss clock-case maker, named Guinand, discovered a process, by which this material, so precious for the arts, can be obtained of very considerable size. His invaluable secret was taken up by an eminent Bavarian artist, Joseph von Fraunhofer,* who possessed more

---

* For an interesting account of this very remarkable man, see the *Conversations-Lexicon*, vol. iv. p. 362 (German.) His monument in Munich bears the appropriate epitaph—"APPROXIMAVIT SIDERA."

means of carrying it into extensive application. He con-
structed two telescopes, the first of ten inches aperture,
the second of twelve ; the former of these is now in the
Dorpat observatory, and to its extreme accuracy we owe
Struve's invaluable observations on the multiple stars.
The latter was purchased for the royal observatory at
Munich, and, in the hands of Professor Lamont, has rendered signal service to the cause of astronomical science.
The premature death of this ingenious optician was a serious interruption to the improvement of the achromatic
telescope. But the processes employed by the Swiss artist
were communicated also to M. Lerebours, a Parisian optician, who has constructed (with object-glasses manufactured by Guinand) telescopes of twelve inches aperture,
and even thirteen-and-a-half. And it is stated by a distinguished writer in the *North British Review*, that M.
Bontemps, of Paris, has acquired the art of making glasses
of three feet diameter, and that Messrs. Chance and Co.,
of Birmingham, have taken out a patent for his process,
and are prepared to construct achromatic lenses of these
enormous dimensions. If this intelligence prove to be
well-founded, we may look for the commencement of a
new era in the science of the heavens.

Having anticipated so much, for the more convenient
distribution of the subject, we come at length to the history of the Reflecting Telescope. The reader will remember, that as the chromatic aberration is entirely the effect
of the *refraction* of light, the easiest and most compendious
mode of getting rid of it altogether, would be to dispense
with refraction, if possible, and adopt the principle of
reflexion in its stead. This principle had been early applied to the microscope, by an Italian priest, Father Zuc-
chi: but the idea of a reflecting telescope was first broach-
ed by James Gregory, in 1663. He failed, however, though
he published a very satisfactory explanation of his plan, to
carry out his views in practice; and before any telescope
on his principle was constructed, Newton had, with his
own hands, and on a different plan, executed, in 1666, a
small reflector, an inch in aperture, and six in length,
which showed the satellites of Jupiter, and the phases of
Venus. This first essay was followed, in 1672, by the
larger instrument, which is still preserved in the library of
the Royal Society.

Although prior in execution, the Newtonian telescope is

posterior in invention to the Gregorian. The latter consists of a tube, open at the extremity nearest to the object, and containing at the other a concave metallic speculum, perforated at the centre.* Directly in front of the aperture of this great speculum, and at a distance a little more than the sum of their focal lengths, is placed a second small mirror, also concave, which is moveable by a screw to a greater or lesser distance from the larger, or as to accommodate itself to the place of the image formed by the first reflexion (which varies with the distance of the object). The image formed in the great speculum is reflected a second time by the lesser one, and thrown to a point just in front of the aperture already described in the great speculum; through which it is viewed in an eye-piece of variable magnifying power, according to circumstances.

It is plain, that a telescope of this construction, though free, of course, from chromatic aberration, must always, when its specula are of a spherical form, be subject, at least, to the lesser evil of spherical aberration; and, indeed, in an aggravated form, as it arises both in the first reflexion in the great speculum, and in the second reflexion in the lesser one. The necessity of counteracting this double imperfection, suggested to a French artist, M. Cassegrain, the idea of the telescope, since called Cassegrainian. In this the great speculum is precisely like Gregory's ; but the small one, instead of being concave, is convex. These opposite curvatures of the specula tend to correct the aberration; for in the one its direction is to the centre, while in the other it is from it : but, as they do not completely neutralize one another, even this construction is affected by spherical aberration, though in a less degree than the Gregorian.

The Newtonian telescope differs from both in several respects. In the first place, its great speculum is not perforated : instead of the small mirror being concave, as in the Gregorian, or convex, as in the Cassegrainian, it is a plane, of an oval form, and inclined at an angle of 45° to the axis of the tube. And thus it reflects the image formed in the great speculum, to the side of the telescope, whence it is viewed with an eye-glass. Newton's original design was, to substitute a glass prism for the

* This speculum is commonly spherical. Were it not for the difficulty of working the surface to a true parabolic figure, the speculum of an astronomical reflector should always be parabolic. The small mirror should be elliptical.

second metallic mirror, in order to save the light which is lost by metallic reflexion. But it has been found so difficult to procure glass perfectly colourless and free from veins, that it is rarely adopted in practice.

This desire of avoiding the loss of light incurred by the second reflexion, suggested a further modification of the reflector, known as the Herschelian telescope, from its illustrious inventor, Sir John Herschel. Like that of the Newtonian telescope, its speculum is not perforated. The peculiarity of its construction is, that the small metal is entirely dispensed with, and the image formed in the large one is viewed directly (in what is called "the front view") by the observer, who sits at the mouth of the tube, and applies the eye-glass directly to the first image. In order that his head may not obstruct too many of the rays coming from the object, the speculum is slightly inclined to the axis of the tube ; and thus the image is reflected not to the centre, as in the other constructions, but to the side at which the observer is situated. We need hardly add, that it is somewhat distorted in consequence.

So much of prefatory history and explanation will not, we trust, be deemed uninteresting or unnecessary. The reader is now, we should hope, in a condition to appreciate the success, which, denied to so many of the most gifted men of past generations, has been reserved to reward the genius, enterprise, and perseverance of an Irish nobleman of our own day.

It will be recollected, that in one particular, at least, the reflecting telescope possesses a decided advantage over the refractor—its freedom from all the embarrassing imperfections of the image caused by chromatic aberration. In order to understand the line of experiment adopted by Lord Rosse for the purpose of improving it, we must now consider a few of its countervailing disadvantages.

In the first place, until Lord Rosse commenced his experiments, it was considered practically impossible to work a true paraboloidal figure in metal specula, at least of any considerable dimensions. Artists, therefore, ordinarily were obliged to content themselves with giving their specula as accurate a spherical figure as possible. Now we need not remind the reader, that a spherical reflector must be affected by aberration as well as a spherical refractor. The cause is precisely the same in both :—the central parts, and those near the circumference, form their respective images at different distances from the face of the mirror, and thus produce indistinctness and confusion.

In the second place, the use of a metallic reflector is, as we have already seen, attended with a greater sacrifice of light than that of a glass refractor. Out of 100 rays received, about 93 are transmitted after refraction; while, according to Herschel, after two reflexions, little more than 45 are available. This is a subject, however, which has never been investigated with sufficient accuracy, and it is now generally believed, that with good speculum metal, carefully and truly polished, the loss of light is by no means so considerable.*

Thirdly, it is a well-known fact, that, even with surfaces of the same dimensions and equal accuracy, the defining power of a reflector is far inferior;—the imperfections of an image obtained by reflexion, being five or six times as great as those of an image formed by refraction.†

Hence it follows, that any one proposing to effect a decided improvement in the reflecting telescope must, 1st, devise some means of counteracting the spherical aberration. 2ndly, he must either improve the reflecting substance, so as to diminish the loss of light; or he must construct specula capable of collecting so much light into the image, as to make this loss a matter of no consequence. 3rdly, he must improve the shape and accuracy of the reflecting surface, in order to avoid the fatal defects of image which attend every inaccuracy in reflexion, however slight it may be.

We pray attention to these principles, while we detail briefly the result of Lord Rosse's long and laborious investigations and experiments. They extend over a period of eighteen years, and have been attended with an expense which none but a princely fortune could have borne.

I. When, in 1826, he commenced his labours, he was well aware of the difficulties by which his path was beset. It was just the time when the hopes of the improvement of the achromatic telescope were at the highest. Discs of very fine glass and of extraordinary dimensions, manufactured by Guinand, of whom we have already spoken, had just been brought into these countries :—one of 12 inches aperture, and 20 feet focal length, by Sir James South ;—

---

* See Dr. Robinson's Address to the Royal Irish Academy " *Proceedings.*" No. 25. p. 7. Nov. 9, 1840.
† Brewster's *Optics*, p. 354.

another of 13¼ inches diameter, and 25 feet focus, by Mr.
Cooper of Marckrea, in the county of Sligo. These were
promising and tempting beginnings in this department.
On the other hand, it required no slight moral courage to
undertake in the reflecting telescope, what, after a long life
spent in its pursuit, such a mind as Herschel's had failed to
accomplish. But little aid was to be looked for from those
who had gone before him in this line. Short, the most
successful speculum maker of his age, had carried his
secret to the grave, and the results of Herschel's extraordinary experience, though almost prepared for publication
before his death, have not even yet been given to the world.
His greatest work, the four-feet speculum, had but a few
years before, been taken down and replaced by one of 18
inches diameter ; and the failure of its polish had begun to
be regarded as an evidence that it was impossible to attain
any considerable dimensions in speculum metal, without
diminishing its brilliancy and durability in such a degree
as to render the instrument comparatively valueless. These
difficulties seem for a time to have had their influence on
Lord Rosse. His first experiments were upon the construction of fluid lenses.* They were (shall we not say happily?)
unsuccessful. Had he succeeded, he might have remained
all his life an ingenious trifler. As it was, he at once
transferred his energies to the more difficult, and certainly
far more honourable line, in which he has since attained
to such distinguished eminence.

His first thought naturally was, how the spherical aberration was to be got over. We may so far anticipate as to
say at once, that he eventually got rid of it, by abandoning
the spherical form altogether. But his first attempt was
so ingenious and interesting, that we cannot pass it by.
The reader will remember that this aberration arises from
the unequal focal length of the centre and of the circumference of a spherical speculum : that of the circumference
and the adjacent parts being shorter than that of the
centre. Lord Rosse endeavoured to make them coincide,
as nearly as possible, by the following ingenious plan.†
Instead of using one unbroken speculum, he divided it into
two concentric parts, or rings, so calculated, that, presu-
ming the original figure to be spherical, each would bear

---

* See Phil. Trans. 1840. p. 503.

† Ibid, p. 521-2; also *Edinburgh Journal of Science*, July, 1828.

about half the aberration of the entire surface. These he
accurately combined and polished as a single surface. In
the original figure, the focus of the centre would, of course,
have been slightly *in advance* of that of the outer ring.
But by a delicate screw-adjustment, the central part was
drawn slightly *back,* and thus its focus was brought into
coincidence with that of the outer part. The aberration
was thus diminished by one-half; and the experiment (with
a telescope of six inches aperture and two feet focus) succeeded so well, that he contemplated a further trial in the
dimensions of eighteen inches aperture, and had already
completed the castings for the purpose, which were in three
concentric portions similarly adjustible. Meanwhile,
however, other lights came upon him, which fortunately
induced him to abandon this ingenious, but hardly practi-
cable course.

The lights to which we allude were such as to satisfy him,
that the prevailing belief, as to the spherical figure's being
the only one securely attainable in polishing a speculum,
was entirely unfounded. A perfectly parabolic speculum
would be entirely free from aberration, for it would reflect
all rays, parallel to its axis, into a single focus; but this
figure it is impossible to attain with complete mathematical
precision. It might be possible, however, Lord Rosse
conceived, to approximate to it; and this approximation
became the subject of a long train of most careful experiment. The extreme delicacy required in this investigation
may be gathered from a fact stated by Dr. Robinson, at
the meeting of the British Association in Cork, that so
slight is the difference between the two figures (the faulty
spherical figure and the perfect parabolic one), that, if two
specula of six feet diameter, the one spherical, the other
parabolic, were in contact at their vertex, the edges would
not diverge from each other more than $\frac{1}{1000}$ of an inch!
Without puzzling our readers or ourselves with any exam-
ination of the mathematical properties of the paraboloid,
we think we shall be able to give a tolerably clear notion
of the principle according to which the approximation to
the parabolical figure, which is obtained by Lord Rosse's
process, has the effect of correcting the aberration.

We have repeated more than once, that this defect is
occasioned in all spherical specula, solely because the image
formed by the central rays is reflected to a greater distance
from the face of the speculum than that formed by the rays

2

reflected from the parts near the circumference. Hence, as the reader will recollect, Lord Rosse's first effort to correct it was, by *drawing back the central part* so as to make the foci of the centre and circumference coincide. Now, the same result would be obtained, were it possible to make the *parts towards the circumference throw their image a little further in advance,* the focus of the centre remaining the same. In the course of his trials with Edwards's polishing-machine, his Lordship found, that when the stroke of the guide of the polisher gives a lateral motion in certain proportions with the diameter of the speculum, the focal length gradually and regularly increases. Hence, remembering that each concentric ring of the entire surface has a separate focal length, it will clearly be possible, provided that of the centre remains the same, and those of the successive rings be gradually increased in the due proportion, to procure an exact coincidence of them all; in the same way in which guns of unequal carrying power may be made to throw their shot to the same point, by regulating the distances from it at which they are severally discharged. Let there be any surface, therefore, originally spherical. A polishing-tool, if it revolve with a radius of uniform length, will clearly have the effect of preserving this original spherical figure. But, if the length of the radius be gradually increased, the abrasion of the metal will become insensibly greater as the tool recedes from the centre; the focal lengths will be gradually increased; and thus the foci of the successive rings of the surface will be brought into coincidence with the central focus of the figure. "According as the focal length increases more or less rapidly, the nature of the curve will vary; and we might conceive it possible, having it in our power completely to control the rate at which the focal length increases, so to proportion the rate of that increase, as to produce a surface approximating to the paraboloid." The most accurate determination of the necessary proportions at which Lord Rosse has yet been able to arrive is, " that, when the stroke of the second eccentric is .27 of the diameter of the speculum, the curve will be nearly parabolic."

It is hardly necessary to say, that the form thus communicated to the speculum practically obviates the first of the three great difficulties against the use of reflectors to which we have alluded. The axis of the paraboloidal figure thus generated, is the axis of the tube. Now, any ray of light impinging on the interior of a paraboloid in a direction parallel to its axis, is reflected accurately to the focus. And so near, in point of fact, has this approximation been brought to the mathematically true figure, " that the three-feet metal at present in the telescope, with its whole aperture, is thrown perceptibly out of focus by a motion of the eye-piece amounting to less than the thirtieth of an inch; and even with a single lens of one-eighth of an inch focus, giving a power of 2592, the dots on a watch dial are still in some degree defined."*

In this success, however, Lord Rosse cannot strictly be said to stand alone. Others, before his time, had succeeded in obtaining a very good parabolic figure, within small dimensions, and in hand-wrought specula. His triumph consists in the invention of a means, by which, with perfect ease and security, the figure can be imparted to specula of *all* dimensions, from one inch to six feet in aperture; and by which what before was always precarious, and at best a labour of days, and even of weeks, may now be accomplished with infallible accuracy, and under the superintendence of a common workman, in a few hours.

II. The polishing and grinding machine, however, would be of limited use in constructing effective reflectors, unless we were able to overcome the second difficulty, which has been explained as inseparable from their use; viz. the enormous sacrifice of light with which it is attended. It is plain, that the quantity of light directly derivable in a telescope from any luminous body, must depend on the aperture. As, therefore, the loss of light in a reflector is far greater than that incurred in a refractor, it would be necessary to obtain with the latter much greater aperture than with the former, in order that their illuminating power should be the same. Unfortunately, the diameter attainable in glass has hitherto been very limited. Dr. Robinson stated at the meeting of the British Association, in 1843, that there is not at present a single object-glass in the world of 16 inches diameter, and not four of 12. And, even with these low dimensions, the cost is enormous. The king of Bavaria's telescope (12 inches aperture) cost £2720. Fraunhofer estimated the expense of one with an aperture of 18 inches, at £9200. For an un-

---

wrought disc of flint glass, 8 inches diameter, and 1 in thickness, about eighty guineas are required; and to this price, we need not say, a very large addition will be made, before an achromatic lens of that diameter can be produced. Hence, it has always been an object of interest with astronomers, to exceed in metal the diameter which Nature seemed to have fixed as the limit attainable in glass. The ordinary speculum metal, however (a composition of copper and tin, and sometimes arsenic or silver, or both together) has been found almost equally intractable. Brittle to a degree perfectly inconceivable to those who have not seen it wrought—liable to all the chances of unequal expansion and contraction, and sensible to an excessive degree of every slight variation of temperature, even those inseparable from the very friction by which it is wrought—it presents almost insuperable difficulties to the artist. A single drop of condensed steam, falling upon a plate of speculum metal, would surely break it; the very pressure of its own weight, in the act of being turned, will destroy its figure; and the application of the polisher, after having been merely washed in hot water, had the effect (though its temperature was hardly altered at all) of cracking plates of considerable thickness and solidity. This excessive brittleness may be, in some degree, overcome by the addition of an extra proportion of copper; but, on the other hand, this expedient dims the brilliancy of the metal, increases its porosity, and materially augments its liability to tarnish.

Under the pressure of all these difficulties, the metal speculum was long kept within dimensions little exceeding those of the achromatic object-glass. The only prominent exception was Herschel, who attained the extraordinary diameter of 48 inches in the celebrated speculum, to which we owe the discovery of the Georgian. But, having had recourse to the plan just alluded to, of increasing the proportion of copper, this speculum was far from realizing the expectations its dimensions would suggest. Its colour was low, and it resisted the effects of tarnish so ill, that, in 1822, it was replaced by a smaller speculum of 18 inches. And thus, since his time, the dimensions of the speculum fell back to the old limit, and even below it: the London opticians do not like to undertake a speculum even of nine inches; and Mr. Grubb, a Dublin artist, stood altogether

alone in the courage to attempt a diameter of fifteen.*
The train of experiment, therefore, by which Lord Rosse
overcame the difficulties which surrounded this portion of
his task, is among the most interesting in the whole history of the chemistry of fusion. They are detailed at great
length by himself, in a paper in the *Philosophical Transactions*, 1840, and an interesting account will be found in
the very excellent pamphlet (No. I.), published at Parsons-town, upon the subject. We can only afford a brief
analysis.

The composition of the speculum metal was his first
study. This had already been the subject of numerous experiments. Newton began with copper and tin in certain
proportions. Edwards tried no less than seventy different
combinations, and in the end preferred the following:—
copper, 32 parts, tin 15, brass 1, silver 1, arsenious acid,
or white oxide, 1. Little discarded silver, as making the
compound too soft: he employed copper 32, tin $16\frac{1}{2}$,
brass pin-wire 4, arsenic $\frac{1}{2}$. Mudge used 32 copper, and
$14\frac{1}{4}$ grain tin: Herschel, 32 copper, 10.7 tin. After a
variety of trials, Lord Rosse found it best to confine himself to the materials used by Newton—"the best proportions being 4 atoms of copper to 1 of tin, or 126.4 parts
of copper to 58.9 of tin."

From the extreme difficulty of managing so brittle a
material, he at first regarded it as impracticable to grind
and polish a *solid* speculum of large dimensions, even
though he could succeed in casting it, which appeared
very problematical: his first thought, therefore, was to
cast a number of small pieces, and to unite them into one
surface, artificially strengthened, which might then be
turned and polished to the required form. There was,
however, a difficulty in the way. To make the speculum
sufficiently strong to bear the operation of polishing, it
would be necessary to unite the plates upon one metallic
frame of sufficient firmness. Now metals, as the reader
is well aware, differ in their conducting power, and are not
equally expanded and contracted by the variations of temperature. If, therefore, the pieces of speculum metal were
united upon a bed of different expansive power from themselves, the unequal expansion and contraction would inevi-
tably crack the speculum metal. It became necessary,

* Dr. Robinson's Address to the Royal Irish Academy. Nov. 9, 1840. p. 3.

therefore, to prepare a metallic bed which would expand
and contract in the same proportion as the speculum
metal. With this view Lord Rosse selected an alloy of
copper and zinc, a species of brass ; and the process by
which he determined the proportions of the alloy is so
curious, that we shall give it in his own words :

"A bar was cast of speculum metal, fifteen inches long and one
inch-and-a-quarter square; smaller bars, but only one-fourth of an
inch thick, were cast of the alloys to be tried, containing a little
more or less than the proportions given, (2.75 of copper to 1 of
zinc.) The two bars were made to fit neatly, so that when brought
together, a very fine line could be drawn across them with hardly
any troublesome parallax at the joint: the whole was then immersed almost to the top, in a tin vessel of water of the temperature of the atmosphere, and that vessel placed in another much
larger also containing water. Pieces of ice were then dropped into
the outer vessel, so that the temperature of the whole was easily and
gradually brought down to nearly 32° : and a straight line, as fine as
possible, was then drawn across both bars, and examined with a microscope, to see that it was perfect. The temperature was then
gradually raised, by pouring hot water into the outer vessel, until
nearly 212° had been attained, and the line was again examined
with a microscope ; and where the alloy had been made by mixing
2.74 of copper with 1 of zinc, and the loss in melting amounted to
$\frac{1}{180}$ of the whole, the continuity of the line was not broken in that
range of temperature ; according, however, as the proportion of the
zinc was more or less, the expansion of the brass bar was greater or
less than that of the speculum metal."*

The difficulty, however, did not cease here. As zinc is
more easily volatilized than copper in the process of fusion,
it was found that in a melting, in which the proper proportions of cold metal had been used, the proportion of zinc
which remained after fusion was very much diminished ;
and, what was still more embarrassing, the amount of zinc
lost was not the same every time, but varied very considerably in different meltings. Although we mention this
as a sample of the endless difficulties of his task, we cannot
enter into the device by which Lord Rosse overcame it,
nor the still more troublesome process by which the pieces
of this metallic bed (eight in number) were attached to
each other. The ordinary process of soldering was found
quite insufficient. Bolts of the same metal as the alloy
would not be sufficiently strong ; while bolts of iron would

* *Phil. Trans.* 1840. p. 506-7.

not contract and expand in the requisite proportions. He was obliged, therefore, to have recourse to the process known among practical men by the name of "burning," in which a stream of molten metal is poured upon the surfaces which it is intended to unite, until they are themselves reduced to a state of fusion at the point where the union is required. This process (which somewhat resembles the re-sealing of a letter which has had the seal broken) was repeated at thirty-two different points, and proved perfectly successful.

Upon this piecemeal plan he constructed three specula; the first 15 inches in aperture, the second 24, and the third 36. The last was in sixteen different plates. The difficulty of casting perfect pieces, even of these small dimensions, led to a great variety of experiments, which we cannot detail, but which resulted most happily in the discovery of a sure mode of casting discs of this material, brittle and intractable as it was believed, of almost any dimensions which may be desired.

He soon discovered that the flaws discernible in the castings, and the excessive brittleness of the whole mass, arose from the metal's contracting unequally at different points before it grew solid after fusion. As the heat is most rapidly conducted away from the edges, these naturally become solid, while the central parts were still fluid. The central mass was thus, as it were, held in a stiff frame during the process of cooling; and, if the surface did not give way in cracks, the parts at least were kept in the most extreme tension, ready to be separated by the slightest violence or the most delicate change of temperature. It occurred to Lord Rosse, therefore, to devise some means of conducting the heat away, gradually and regularly, from the lower surface, so that the upper one might retain it in a greater proportion. The result would evidently be, that the metal would become solid first in its lowest stratum, from which the solidification would proceed regularly upwards through the successive strata, so as to prevent both the strains and flaws which were unavoidable in the old process. He tried two different modes of effecting this.

First, he made a mould of cast-iron, and cooled the lower surface by a constant jet of cold water. This proved the correctness of the principle; but the mould uniformly cracked during the process. He next tried a mould partly made in the usual way, the sides being of packed sand; but, for the bottom of the mould, he used instead of sand a plate of iron: the result was, a casting without cracks, but having its surface full of flaws and air-bubbles. The reason of this defect evidently was, that the superior conducting power of the iron bottom drew off the heat so rapidly from the lower surface, that the disengaged air could not make its way through the solid bottom of the mould; and this failure suggested the crowning idea of the whole process, which was to form the bottom of the mould of iron hoop-plates placed vertically against each other, so close as to retain the fluid metal, but yet sufficiently free to allow the air to escape without impediment. This very simple, but ingenious plan, has proved thoroughly successful. Of sixteen plates cast for the three-feet speculum, not a single one failed, and on every after occasion, though he enlarged the diameter to 20 inches, and even 6 feet, the castings were always perfect at the very first trial.

Of the process of annealing we need hardly speak. It resembles very much that pursued in glass manufactories, allowing for the difference of material. The time of cooling, of course, varies with the size of the mass. For a nine-inch plate it is three or four days; the great six-feet speculum was allowed to remain in the annealing oven sixteen weeks; but a second speculum of the same dimensions, was perfectly annealed in three weeks.

Here, then, is another advance in the construction of reflecting telescopes, the importance of which it is almost impossible as yet to estimate. A speculum with such enormous aperture is so immeasurably beyond the dimensions attainable in glass, that henceforward the greater loss of light, consequent on the use of reflectors, need hardly be taken into the calculation of their respective perfectibility.*

III. Still, however, there remained a third difficulty to which Lord Rosse was now obliged to address himself. It was not enough to obtain great illuminating power, and entirely free from spherical aberration. It still remained to remove, or at least abate, another defect, which seemed inseparable from the use of metallic reflectors—want of defining power. It is ascertained by careful experiment,

---

* The light received is as the square of the diameter of the aperture. Hence, Lord Rosse's *six-feet* speculum will receive *thirty-six* times as much light as a *twelve-inch* achromatic.

that the imperfections of an image formed by a reflector are five or six times greater than those of a refractor, of the same dimensions and the same degree of accuracy. The next step, therefore, was to procure a means of communicating a perfectly true figure and fine polish.

This result is obtained with unerring accuracy, by a modification in the use of the machinery already referred to in describing the mode of obtaining a parabolic figure. Some notion may be formed of the extreme accuracy required, from a statement of Lord Rosse, that an error of a small fraction of a hair's-breadth would destroy all hope of correct action; and Dr. Robinson states, that the smallest inequality of local pressure during the polishing process would "change a well-defined star into a blot or a comet." The old process of polishing (which up to a diameter of nine inches was performed with the hand, and, beyond that dimension, required that at least the cross-stroke should be given with a lever moved by the hand) was liable to three capital difficulties: 1st. variations of the extent and velocity of the motions requisite in the operation; 2nd. variations of the temperature of the speculum and of the tool; 3rd. variations of pressure both at different times and at different points of the surface. The surest, and indeed the only sure way, to obviate these, was by the use of machinery, which should do its work with certainty and regularity, and have a tendency to correct its own defects. Accordingly, Lord Rosse's machinery (which we had the good fortune to see while in operation upon the six-feet speculum) produces the necessary shape and polish with the most unerring accuracy. The speculum, with its face upwards, is made to revolve slowly, immersed to within an inch of its surface in a cistern of water, regulated to a temperature of 55° Fahrenheit. The polishing-tool is drawn longitudinally along its surface by the stroke of an eccentric, which is adjustible to any stroke, from 0 to 18 inches; and it receives at the same time a transverse motion from another eccentric similarly adjustible. During a single revolution of the speculum, the polisher makes longitudinally thirty-seven strokes (each one-third of the diameter of the speculum), and transversely 1.76 (each about .27 of the same diameter). Though it has no direct rotatory motion, yet it is carried round by the revolution of the speculum, once for every fifteen or twenty revolutions of the latter, and its pressure is regulated by a

counterpoise, so as to be uniformly about 1℔ for every
superficial circular foot of the speculum. We need hardly
say, that this arrangement obviates all the irregularities
detailed above. It secures perfect regularity of action,
places the temperature completely under control, and regu-
lates the pressure with unerring uniformity. The material
used in *grinding* the speculum is emery and water, and
the tool is intersected by longitudinal, transverse, and
circular grooves. For *polishing* the movements are the
same; but, as the material employed is different, and as
the little minutiæ of the process are a good example of the
extreme patience and ingenuity of the noble inventor, we
shall give the description in the words of his address to
the British Association:

" The process of polishing differs very essentially from that of
grinding : in the latter, the powder employed runs loose between
two hard surfaces, and may produce scratches possibly equal in depth
to the size of the particles : in the polishing process the case is very
different ; there the particles of the powder lodge in the comparatively soft material of which the surface of the polishing-tool is
formed, and as the portions projecting may bear a very small proportion to the size of the particles themselves, the scratches necessarily will be diminished in the same proportion. The particles
are forced thus to imbed themselves, in consequence of the extreme
accuracy of contact between the surface of the polisher and the
speculum. But as soon as this accurate contact ceases, the polishing process becomes but fine grinding. It is absolutely necessary,
therefore, to secure this accuracy of contact during the whole
process. If the surface of a polisher, of considerable dimensions,
is covered with a thin coat of pitch, of sufficient hardness to
polish a true surface, however accurately it may fit the speculum, it will very soon cease to do so, and the operation will fail.
The reason is this, that particles of the polishing-powder and
abraded matter will collect in one place more than another, and as
the pitch is not elastic, close contact throughout the surfaces will
cease. By employing a coat of pitch, thicker in proportion as the
diameter of the speculum is greater, there will be room for lateral
expansion, and the prominence can, therefore, subside, and accurate
contact still continue ; however, accuracy of figure is thus, to a
considerable extent, sacrificed. By thoroughly grooving a surface
of pitch, provision may be made for lateral expansion contiguous to
the spot where the undue collection of polishing-powder may have
taken place. But, in practice, such grooves are inconvenient, being
constantly liable to fill up : this evil is entirely obviated by grooving
the polisher itself, and the smaller the portions of continuous
surface, the thinner may be the stratum of pitch.

" There is another condition, which is also important, that the pitchy surface should be so hard as not to yield and abrade the softer portions of the metal faster than the harder. When the pitchy surface is unduly soft, this defect is carried so far, that even the structure of the metal is made apparent. While, therefore, it is essential that the surface in contact with the speculum should be as hard as possible, consistent with its retaining the polishing-powder, it is proper that there should be a yielding where necessary, or contact would not be preserved. Both conditions can be satisfied by forming the surface of two layers of resinous matter of different degrees of hardness ; the first may be of common pitch, adjusted to the proper consistence by the addition of spirits of turpentine, or rosin ; and the other I prefer making of rosin, spirits of turpentine, and wheat flour, as hard as possible, consistent with its holding the polishing-powder. The thickness of each layer need not be more than one-fortieth of an inch, provided no portion of continuous surface exceeds half an inch in diameter: the hard resinous compound, after it has been thoroughly fused, can be reduced to powder, and thus easily applied to the polisher, and incorporated with the subjacent layer by instantaneous exposure to flame. A speculum of three-feet diameter thus polished, has resolved several of the nebulæ, and in a considerable proportion of the others has shown new stars, or some other new feature."

The device by which he contrived, at any point of the process, to ascertain the progress of the polishing, is extremely simple. The polishing-machine is situated on the ground floor of a high tower, on the top of which is erected a mast of such height, that its summit is ninety feet from the speculum which is being polished. To this the dial-plate of a watch is attached. A small plane metal is so arranged, that, along with the great speculum in the machine, it forms, for the time being, a Newtonian telescope, furnished with the regular eye-pieces. The dial-plate is the object of this *provisional* telescope ; and, at any moment of the process, there needs but to draw the tool aside, and open a series of trap-doors in the tower, in order to have a trial of the condition of the speculum, if not final, at least abundantly sufficient for every practical purpose.*

The success of this new process in increasing the defining power of the reflector can hardly be conceived. For those who have had no experience in the use of the astronomical telescope, it may be well to observe, that a certain propor-

---

* This is only one of the advantages of polishing with the face of the speculum upwards. It obviates all danger of accidental injury to the speculum.

tion* must always be observed between the magnifying power of the eye-glass and the greater or less perfection —technically called "sharpness of definition"—of the image. If the image be ill-defined, from whatever source, whether from the real imperfection of the object-glass or the speculum, or from the state of the atmosphere and want of light at the time, it is impossible to use an eye-glass of high magnifying power; for this would destroy the apparent distinctness, which, when seen with low power, it is thought to possess. The reason, of course, is, that, by enlarging the image, the greater magnifying power diminishes its brilliancy and detects and displays its imperfections more than the lesser; and as the result of these, whatever may be their cause, is indistinctness, the use of powerful glasses will frequently change a tolerably well-defined image into a shapeless blot. Hence, the best test of a true surface, *cæteris paribus*, is the capacity of, as it is called, "bearing a high magnifying power." As an evidence of the superiority of Lord Rosse's polishing-tool, in every little detail, over that ordinarily adopted, it will be sufficient to mention, that in the polishing of the three-feet speculum, the mere substitution of grooves in the surface of the tool, for the old plan of grooving the surface of the pitch, had such an effect, that it defined better with a magnifying power of 1200, than it had done, when polished in the old way, under a power of only 200.

Having thus detailed the several processes by which the specula are cast to any required size, ground to a parabolic figure, and polished to the exquisite degree of accuracy— technically called a "black polish"—it remains to explain the construction of the telescopes in which the three-feet speculum, and the still more enormous one of six feet, have been mounted.

It will be remembered, that at the time when the new light broke upon Lord Rosse, with reference to the process of casting speculum metal, he was engaged in preparing the compound or plated speculum three feet in diameter, which has been already described. The reflecting surface thus obtained was the largest ever mounted on any telescope, except the great four-feet speculum of Sir William Herschel. The performance of this interesting instrument

---

* This proportion will be found explained very satisfactorily in the *Encycl. Britannica.* (7th Ed.) Art. *Telescope.*

PLATE I.

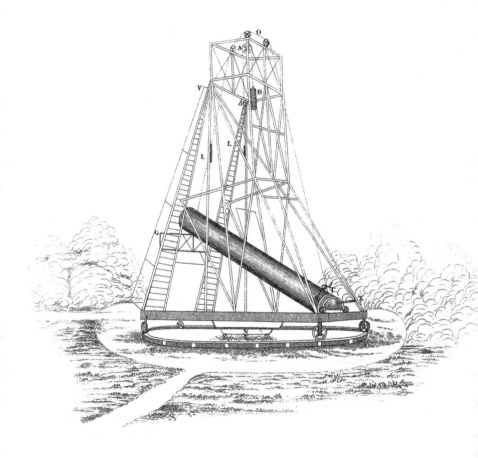

Machinery of the 3 feet Reflector.

has, in some respects, been very satisfactory. For stars below the fifth magnitude it is not at all inferior to the solid three-feet speculum; but Dr. Robinson states, that in all stars "above that magnitude, it exhibits a cross formed by the diffraction at the joints; and, in unsteady states of the air, it exhibits the sixteen divisions of the great mirror on the star's disc."

The great objects of interest, therefore, are the solid specula, the first of three, the second of six feet aperture. We shall take them in order.

The polishing of the three-feet speculum, at which several eminent scientific men, among others Dr. Robinson, assisted, was perfected in the short space of six hours. Its mounting has but little of peculiarity. It is fixed in a box, in which, to obviate the possibility of strain or displacement from any flexure of the wood, it is supported on nine plates of iron, each a sector of a circle of the same dimensions with the speculum itself. The plates "rest, at their respective centres of gravity, on points supported by levers, which rest on three original points,"* the lever apparatus being "exceedingly, and indeed disproportionately, substantial; otherwise tremors would be introduced by it, attended with the worst consequences."

In all other details the mounting is very similar to that of Herschel's telescope, except that the tube and gallery are both counterpoised, in consequence of their very great weight. The tube is twenty-six feet long, but the machinery by which its movements are directed is so admirably contrived, that it is managed with the utmost facility. The accompanying plate† (No. I.) will enable the reader to understand its construction. A graduated iron circle, thirty feet in diameter, is fixed in the ground, and serves as a rail, on which four grooved wheels revolve freely, supporting the entire structure, and enabling the telescope to be directed to any quarter of the heavens. The centre upon which the whole frame moves is a pivot, passing through the beam marked N, which is firmly bolted to the great cross-beam immediately above it in the engraving. By means of a lever, on the long arm of which a weight is made to act with any required amount

---

* It is stated differently in " The Monster Telescopes," but we, of course, follow Lord Rosse's own account in the *Phil. Trans.* 524.

† Copied, as are the two following, from the very excellent illustrations of " *The Monster Telescopes.*"

of pressure, while its shorter arm presses against the end of the beam N, the entire structure may be counterpoised at pleasure. The lower end of the tube is free; and, when the other extremity is raised or depressed, the axle (T) runs easily on wheels (seen in the engraving) which slide with the utmost facility upon a railway traversing the great main-beam. The counterpoise of the tube is seen at B upon the figure. The gallery on which the observer stands, is counterpoised by the weight L, and rises and falls with the tube. When it is necessary to raise or lower the latter, it is done with the utmost facility by the small wheel and axle T, from which a rope may be seen to run across the top pulley O, whence it descends to the mouth of the tube; and by a very simple arrangement the observer is enabled, without leaving the box, to raise or depress it for short distances. The speculum-case can be opened at pleasure, and a box filled with quick-lime is kept constantly within it, for the purpose of absorbing the moisture and acid vapours from which the speculum would otherwise suffer.

The telescope is of the Newtonian construction. It is furnished with a variety of eye-pieces, of different powers, from 180 upwards. The power to be employed depends mainly on the state of the atmosphere, and this varies very much in our climate. The three-feet speculum will bear a power of 2,000 on some nights, better than one of 100 on the night preceding. The powers employed with it, there-fore, vary from 180 to 2,000, and sometimes, but very rarely, go higher. From some observations in the paper so often cited from the Philosophical Transactions, it would appear that Lord Rosse has tried it as a Herschelian (dispensing with the second metal); but, although the distortion of the image was not so great as he expected, yet the saving of light by no means compensated, in a telescope of such enormous aperture, for the sacrifice of distinctness (occa-sioned by the oblique reflexion) at which it was purchased.

Instead, however, of speculating upon the probable capa-city of the instrument, it will be more interesting to give a short report of its actual performance, as described in the Royal Irish Academy by Dr. Robinson, who, from the beginning, had watched the progress of the experiments with the utmost anxiety. The trial which formed the subject of this Report, was made more for the purpose of ascertaining the *defining* power of the telescope, than of eliciting any new astronomical fact: and it will be well to bear this in mind, in order to understand the tenor and tendency of some of Dr. Robinson's observations.

"In trying the performance of the telescope, Dr. R. had the advantage of the assistance of one of the most celebrated of British astronomers, Sir James South; but they were unfortunate in respect to weather, as the air was unsteady in almost every instance; the moonlight was also powerful on most of the nights when they were using it. After midnight, too (when large reflectors act best), the sky, in general, became overcast. The time was from October 29th to November 8th.

"Both specula, the divided and the solid, seem exactly parabolic, there being no sensible difference in the focal adjustment of the eyepiece with the whole aperture of thirty-six inches, or one of twelve; in the former case there is more flutter, but apparently no difference in definition, and the eye-piece comes to its place of adjustment very sharply.

"The solid speculum showed α Lyræ round and well defined, with powers up to 1600 inclusive, and at moments even with 1600; but the air was not fit for so high a power on any telescope. Rigel, two hours from the meridian, with 600, was round, the field quite dark, the companion separated by more than a diameter of the star from its light, and so brilliant that it would certainly be visible long before sunset.

"ζ Orionis, well defined, with all the powers from 200 to 1000, with the latter a wide black separation between the stars; 32 Orionis and 31 Canis minoris were also well separated.

"It is scarcely possible to preserve the necessary sobriety of language, in speaking of the moon's appearance with this instrument, which discovers a multitude of new objects at every point of its surface. Among these may be named a mountainous tract near Ptolemy, every ridge of which is dotted with extremely minute craters, and two black parallel stripes in the bottom of Aristarchus.

"The Georgian was the only planet visible; its disc did not show any trace of a ring. As to its satellites, it is difficult to pronounce whether the luminous points seen near it are satellites or stars, without micrometer measures. On October 29, three such points were seen within a few seconds of the planet, which were not visible on November 5; but then two others were to be traced, one of which could not have been overlooked in the first instance, had it been in the same position. If these were satellites, as is not improbable, there would be no great difficulty in taking good measurement both of their distance and position.

"There could be little doubt of the high illuminating power of such a telescope, yet an example or two may be desirable. Between ϵ^1 and ϵ^2 Lyræ, there are two faint stars, which Sir J. Herschel (Phil. Trans. 1824) calls 'debilissima,' and which seem to have been, at that time, the only set visible in the twenty-feet reflector. These, at the altitude of 18°, were visible without an eye-glass, and also when the aperture was contracted to twelve inches.   With an aperture of eighteen inches, power 600, they and two other stars (seen in Mr. Cooper's achromatic of 13.2 aperture, and the Armagh reflector of 15) are easily seen.   With the whole aperture, a fifth is visible, which Dr. R. had not before noticed.   Nov. 5th, strong moonlight.

" In the nebula of Orion, the fifth star of the trapezium is easily seen with either speculum, even when the aperture is contracted to eighteen inches.   The divided speculum will not show the sixth with the whole aperture, on account of that sort of disintegration of large stars already noticed, but does, in favourable moments, when contracted to eighteen inches.   With the solid mirror and whole aperture, it stands out conspicuously under all the powers up to 1000, and even with eighteen inches is not likely to be overlooked.

" Comparatively little attention was paid to nebulæ and clusters, from the moonlight, and the superior importance of ascertaining the telescope's defining power.   Of the few examined were 13 Messier, in which the central mass of stars was more distinctly separated, and the stars themselves larger than had been antici-pated; the great nebula of Orion and that of Andromeda showed no appearance of resolution, but the small nebula near the latter is clearly resolvable.   This is also the case with the ring nebula of Lyra; indeed, Dr. R. thought it was resolved at its minor axis; the fainter nebulous matter which fills it is irregularly distributed, having several stripes or wisps in it, and there are four stars near it, besides the one figured by Sir John Herschel in his catalogue of nebulæ.   It is also worthy of notice, that this nebula, instead of that regular outline which he has there given it, is fringed with appendages, branching out into the surrounding space, like those of 13 Messier, and in particular, having prolongations brighter than the others in the direction of the major axis, longer than the ring's breadth.   A still greater difference is found in 1 Messier, described by Sir John Herschel, as 'a barely resolvable cluster,' and drawn, fig. 81, with a fair elliptic boundary.   This telescope, however, shows the stars, as in his figure 89, and some more plainly, while the general outline, besides being irregular and fringed with appen-dages, has a deep bifurcation to the south."

Such were the results of the first trial of the three-feet reflector.   Being intended, as we have already observed, chiefly to test the defining power of the instrument, it falls far short in interest of the subsequent trials made on every occasion which offered by Lord Rosse himself.   We shall return to these before the close : we must first say a word upon the six-feet reflector.

Even before Lord Rosse had completed the instrument

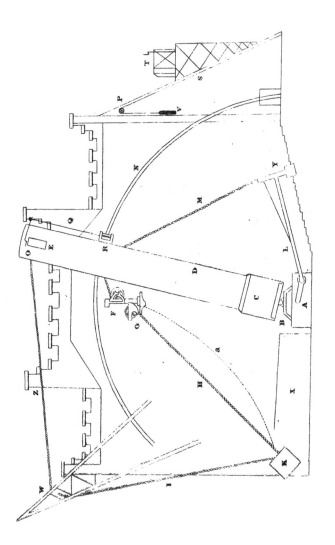

PLATE II.

Section of the Machinery of the 6 feet Reflector.

just described, he had projected another, of the still more extraordinary dimensions of six feet. The project has since been carried into execution, at an expenditure, it is said, of £12000. The arrangements for the construction of the smaller speculum, required only to have their scale enlarged in order to be perfectly applicable to the greater. In consequence of the greater mass of metal (above three tons) it was found necessary to construct a foundry for the special purpose of casting it, and by using three furnaces, from which the crucibles were conveyed simultaneously by machinery, and poured at the same moment into the mould, the casting was made perfectly uniform and regular.* At the first appearance of incipient solidification, it was carried upon a railway to the annealing oven, where it was left for sixteen weeks. It is supported in the speculum-box on a plan similar to that already described as used for the three-feet speculum, but more complicated; the bed consisting of twenty-seven, instead of nine plates.† In principle, however, the support is precisely the same, and is admirably contrived to obviate the possibility of flexure in any direction; whether lateral, horizontal, or oblique. The polishing of the speculum was completed in six hours so successfully, that, although it was not intended to be final, yet the performance of the instrument, when tried, gave such satisfaction, as to induce Lord Rosse to leave it in its present condition.

The tube, with the speculum-box, is fifty-six feet in length, and seven feet in diameter. The box is so capacious as freely to admit two men, for the purpose of removing or replacing the cover; it is connected with the tube, however, in such a way as not to exert the slightest pressure upon it. As its movements, in consequence of its immensely greater weight and size, are much more limited than those of the former instrument, it may be as well to say a word in explanation of its machinery. The tube is fixed by an enormous universal joint (marked B, Plate II.), like that of a pair of compasses moving round a pin; so as to have not only a vertical, but also a transverse motion, for the purpose of following the object in right ascension. It stands in the centre, between two castellated walls, parallel to the meridian, 72 feet long, 56

---

* To secure uniformity, the metal was first melted in three crucibles, and each of these meltings being broken into pieces, the pieces of each melting were put successively into the crucibles for the final casting.

† These plates are lined with pitch and felt, to guard (by their non-conducting power) against variation of temperature.

high, and 24 feet asunder.   The tube, therefore, ranges
north and south : in the latter direction it may be lowered
to the horizon; in the former it is, of course, sufficient to
bring it to the pole.   The lateral movement is necessarily
limited by the twenty-four-feet distance between the walls;
but this gives half-an-hour on each side of the meridian,
which is fully sufficient for all the purposes to which such a
telescope is likely to be applied.   The observer's gallery is
detached from the tube, to avoid tremors; and it is, of
course, needless to say that both are counterpoised.   The
first* plan of the counterpoise was a chain passing from the
end of the tube over a pulley, and carrying the counterpoise, which was to run on a curved railway, so formed that
the telescope should be in equilibrium through its entire
range.   But, though the preparations for this plan were all
made, it has been effected in a much more simple manner ;
which is remarkably well described in the account of the
telescopes to which we have so often referred.   We regret
that we must be content with briefly stating the principle.
The counterpoising weight, instead of rolling upon the
curved railway, as originally intended, is attached by a
guy to a fixed point, by which it is made to describe a
curve in its ascent and descent, so that when the tube
is low (and, therefore, acts with greater pressure), the counterpoise acts at a proportionally greater mechanical advantage ; and when the tube is high, its advantage propor-
tionably diminishes.

The transverse motion is effected with equal facility.
An iron circle is fixed to the eastern wall, and from this
runs a wooden pole, through the iron bed, which is fastened to the tube.   A rachet is fixed to the iron circle, and a
handle acts upon this by means of an endless screw.   This
is, at present, worked by the hand, but it is intended to
move by clock machinery, adjusted to the required rate
of motion.

As the observer's gallery, however, is independent of
the tube, there is a distinct provision by which he is enabled to follow all the movements described above.   Within
certain limits, as far as the ladders, shown in the front of
the engraving (Plate III.), permit, the gallery is elevated
by a windlass, fixed at the side of the wall ; and the observer's box is moved with the utmost ease across the
gallery, by a handle wrought by the observer himself.
Beyond this point he follows the tube on a series of three

---

* *North British Review.* No. iii. p. 208.

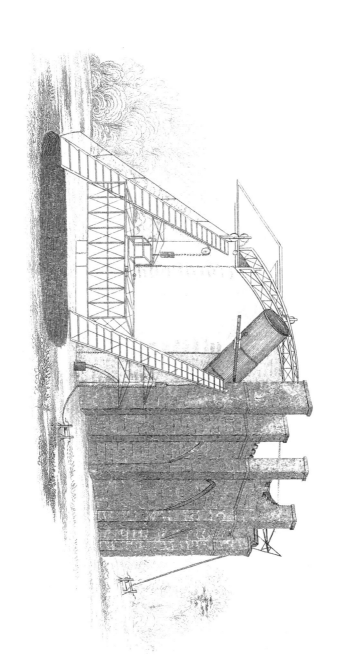

PLATE III.

Perspective view of the 6 feet Reflector.

galleries, of a very solid and ingenious construction, which may be wheeled out in succession, so as to reach the greatest point of distance in the transverse range of the tube.

All these movements are effected at present by the hand, but arrangements, we understand, are in progress, by which a motion, regulated by the movements of the object (as in the equatorial), can be imparted by machinery. This telescope, as well as the three-feet one, is used as a Newtonian, the small plane metal being placed near the mouth of the tube at an angle of 45° to the axis. It is so contrived, that, if deemed advisable, it may be used as a Herschelian; but with an aperture so enormous, the saving of light can hardly be an object of so much importance as to warrant the risk of indistinctness;* and whether it shall ever be so used remains, we are informed, to be determined by future experiments. What amount of magnifying power in the eye-glass it may bear under the most favourable circumstances, cannot as yet be determined. The lowest power which will produce an emergent pencil of $\frac{2}{10}$, (the largest the eye will admit) is 360. Therefore, 360 is the lowest power which will render the whole aperture effective. And, hence, whenever the atmosphere will not bear a power of 360, there can be little use in employing this telescope. It is at present provided with eye-pieces of all focal lengths from 4 inches to $\frac{1}{20}$ of an inch.

We feel that all these dry and minute details will be tedious and uninteresting to many of our readers. But we have hardly thought ourselves at liberty to omit any portion; partly, because the subject is one of great national interest, but still more, because it may be useful to direct attention even to the merest minutiæ of a process so pregnant with important results to the cause of science. It remains for us to say a word upon a more interesting topic —what additions have been made, or may be expected to be made, by these instruments, to the existing stock of positive knowledge of the heavens. The trial which they have received, especially the six-feet speculum, is not sufficient to warrant a conclusive judgment as to minute facts. The great telescope, owing to the absence of Lord Rosse from home, has been but once directed to the heavens.† The

---

* In his paper in the *Phil. Trans*, 1840, Lord Rosse states, that he was then engaged in experiments, as to the practicability of constructing a speculum which should form a portion of a paraboloid *whose axis coincided with the side of the tube.* This would give at the mouth of the tube an image entirely free from distortion.

† Written in the beginning of January.

object examined upon that occasion was a nebula, selected
as an example for the Royal Society, one of a class the
least promising, and least likely to yield new truths under
increased optical power. It is known to be a cluster, and
little more is likely ever to be known regarding it, unless
accurate measurements should hereafter detect motion.
With the appearance of this cluster in the three-feet reflector, Lord Rosse was perfectly familiar; and the contrast of its appearance in the large one was most striking.
But, before he had time to observe it fully the night
changed, and as the sky became overcast before any other
remarkable object came within view, he saw nothing more
of any interest.

The observations with the three-feet reflector have been
more numerous, and a detailed account of some of them
was communicated to the Royal Society on June 13, accompanied by drawings of five of the nebulæ of Sir John
Herschel's catalogue, as seen in this telescope. Before we
transcribe the description of these observations, we think
it right to offer a word of caution as to the *sort* of performance which should be expected from this instrument. In
the first place, it may not be unnecessary to remind some of
our readers, that when we speak of a telescope's manifying
100 or 200 times, we refer not to the apparent diameter of
the object as seen in the telescope, but to the magnitude of
the angle under which it is seen. Now the increase or di-
minution of the apparent size, does not suppose any variation in this angle; but depends entirely (the angle being
given) on the distance of the object. And, hence, in consequence of the immense distance of the fixed stars, their
apparent magnitude is never increased, no matter how
powerful the instrument, nor are their *real* discs rendered
sensible. The only effect of instrumental power upon
them, is to increase their *brilliancy;* and this is increased
in the ratio of the aperture of the telescope, or as the square
of its diameter. In the next place, it must be borne in
mind, that, strange as it may seem, Lord Rosse's telescopes,
though they may of course be used, have not been designed
for the study of the bodies in the solar system. At the
time when he commenced the task of their construction, he
conceived that the phenomena of our system had been sufficiently investigated, with the exception, perhaps, of Saturn's ring, and the anomalous motions of the satellites of
the Georgian; and for these he conceived that existing
instrumental power was abundantly sufficient. To the conditions necessary, therefore, for this branch of observation

—nice contrivances for measurement, and for the compaparison of masses and distances—he paid no attention. His ambition was turned rather to those vast "varieties of untried being," which, with the exception of the Herschels, had till then received but little attention from practical men ;—the worlds and systems of worlds known under the common name of multiple stars, and still more the mysterious luminous masses called nebulæ, every detail of which opens a new problem in cosmogony, and suggests matter for almost limitless speculation. It is not difficult to understand, that the requirements of an instrument which might hopefully address itself to this sublime investigation, must be entirely peculiar. They must evidently be, great magnifying power, and, still more, great capacity for light. If, as astronomers teach, these faint patches of light be in reality systems of distant worlds, vast aggregates of separate luminosities, so remote as to form but one confused streak of blended light, how vast must be the capacity of collecting rays, and how perfect the power of condensing them, which would suffice to resolve into their numberless distinct and discernible images these faint blotches of nebulosity, barely traceable upon the dark surface of the sky even with the most powerful instruments! How great, and, still more, how accurate, the magnifying power which could bring even the minutest image of each under the eye ! These are the objects which Lord Rosse's telescopes were intended to effect, and, of course, it is by this test they are to be tried.

Let us, therefore, take their effect upon the nebulæ as an illustration of their power. There are few who have not heard and read of these mysterious masses of light, of an assemblage of which the Milky Way is a familiar example. But there are many who may not be aware of their number and extent. It is well ascertained, that the number of nebulæ whose places are known in our hemisphere, is nearly two thousand—all nebulous to the naked eye—but when viewed in the telescope, some presenting the appearance of minute, but yet clearly distinct stars; others showing the stars distinct, but so close as almost to blend into one mass of faint light; others, again, not inaptly described as patches of star-dust flung irregularly upon the sky; others, in fine, losing even this shadow of individuality, and blent faintly and dimly into little streaks of misty light. From the time when the telescope first resolved the less distant of these extraordinary masses, and revealed the fact, that these at least were but assemblages of distant stars, this had been the generally received theory about all nebulæ ; and it was believed, that even those which still remained unresolved, differed from the rest only in being incomparably more distant ; and that their ultimate resolution was reserved as a triumph of the still greater perfection of telescopic power. But some of Herschel's observations have suggested the idea, that certain of them, at least, are not collections of real stars ; but rather what they seem to be, huge masses of subtle and attenuated luminous matter—the material, perhaps, from which suns and systems may hereafter · be elaborated, but, as yet, floating in the infinite regions of space, unorganized, or at least still under the process of organization. There are certain circumstances connected with some of them, which give apparent probability to this theory. We may instance two very remarkable ones: the first in the Sword of Orion, the second in the Girdle of Andromeda. Of other nebulæ it is found that, though to the naked eye and to instruments of low power they present the nebulous appearance, yet they lose it under the scrutiny of higher optical power. But of these two it is very remarkable, that, though *visible as nebulæ to the naked eye,* and though under each successive accession of instrumental power they become more brilliant, yet *even to the finest instruments they give no appearance of resolvability.* This, it must be confessed, is a sort of evidence, that these bodies must be nebulous really and in truth, and not alone from their immense distance.

Upon these and similar grounds is built what is called the Nebular Hypothesis, propounded by La Place.* The nebulous matter diffused through space in this, and even far less palpable forms, is,† according to his hypothesis, the material from which all the great bodies of the visible creation are composed. All the nebulæ are observed to have certain centres, more dense and brilliant than the rest. Around these denser nuclei is gradually collected the whole mass of matter within the sphere of their attraction: a process of gradual condensation thus goes on: the nebulous mass increases in solidity, and, consequently, diminishes in diffusion. Rotation commences ; and, by a fixed and well-known law, increases with the condensation of the mass. The body at last acquires a certain solidity, of

---

* Systeme du Monde, ii. p. 418.   Part of the chapter is translated by Nichol—*Architecture of the Heavens,* p. 204—but it deserves to be read entire.

† For example, the *Resisting Medium,* the existence of which the variation in the time of Encke's comet seems to place beyond question.

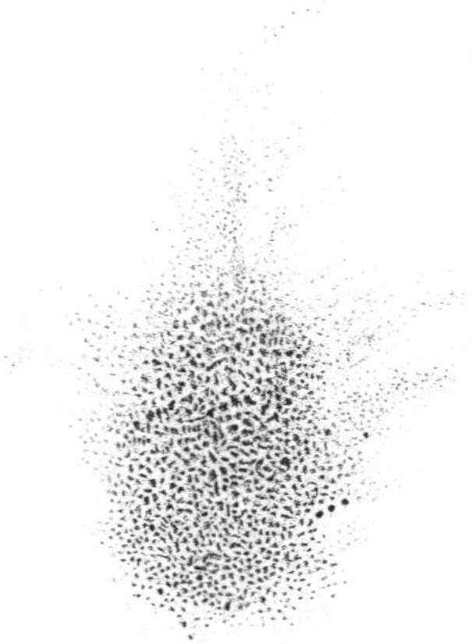

R.A. 19ʰ 52′
Dec. 32° 49′ North.

Fig. 26.

R A. 5ʰ 24′
Dec. 21° 53′ North.

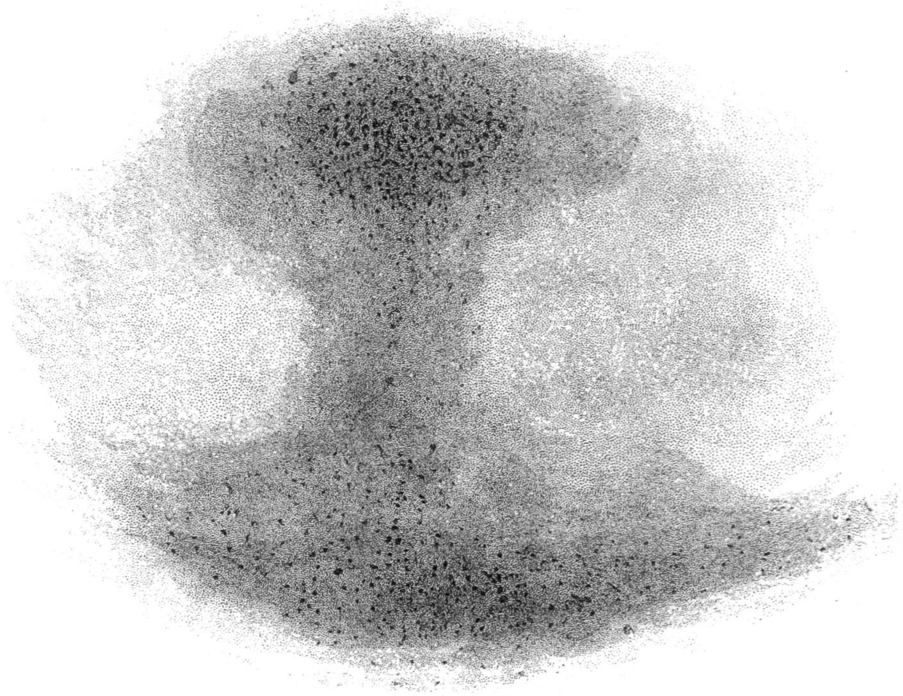

Fig. 81.

Nebulæ as seen in Lord Rosse's 3 feet Reflector.

which, perhaps, the comets of our system are an example, and eventually becomes a planet, or even a sun itself, the centre of a system.

One of the chief grounds of this daring hypothesis is, the supposed *globular form* of the vast majority of the known nebulæ.* Now, this assumption appears to be entirely disproved by Lord Rosse's observations with the three-feet telescope, and it is, of course, to be presumed, that the further instrumental power of the large telescope will reveal still more of the mystery. We subjoin his Lordship's analysis of these observations on five of Herschel's nebulæ:

"Plate XVIII. FIG. 88, is one of the many well-known clusters; I have selected it merely for the purpose of showing that in such objects we find no new feature, nothing which had not been seen with instruments of inferior power; the stars, of course, are more brilliant, more separated, and more numerous. I fear that no amount of optical power will make these objects better known to us, though perhaps exact measurements may bring out something.

"FIG. 81 is also a cluster; we perceive in this, however, a considerable change of appearance; it is no longer an oval resolvable nebula; we see resolvable filaments singularly disposed, springing principally from its southern extremity, and not, as is usual, in clusters, irregularly in all directions. Probably greater power would bring out other filaments, and that it would then assume the ordinary form of a cluster. It is studded with stars, mixed, however, with a nebulosity, probably consisting of stars too minute to be recognized. It is an easy object, and I have shown it to many, and all have been at once struck with its remarkable aspect. Every thing in the sketch can be seen under moderately favourable circumstances.

Plate XIX. FIG. 26, on the contrary, is a difficult object; it requires an extremely fine night, and a tolerably high power; it is then seen to consist of innumerable stars, mixed with nebulosity; and when we turn the eye from the telescope to the Milky Way, the similarity is so striking, that it is impossible not to feel a pretty strong conviction, that the nebulosity in both proceeds from the same cause.

"FIG. 29.—The annular nebula in Lyra; 2 is the star in Sir John Herschel's sketch; I have inserted the six other stars as, in some degree, tests of the power of a telescope. Near star 3 there are two very minute stars seen with great difficulty; the others are easily seen, whenever the light is sufficiently good to show the nebulæ well. The filaments proceeding from the edge become more conspicuous, under increased magnifying power within certain limits, which is strikingly characteristic of a cluster; still I do not feel confident that it is resolvable. I am, however, disposed to think, that it was never examined when the instrument was in as

* See Nichol's Architecture of the Heavens, p. 130.

good order, and the night as favourable, as on the several occasions when the resolvability of figure 26 was ascertained.

" Fɪɢ. 47 is one apparently of another class. It has a star in the centre, and is of unequal brightness ; the nebulosity is in patches, and I have sometimes fancied, though probably erroneously, that I could discover in it a faint resemblance to figure 26. The star in the centre is easily seen, and there is nothing peculiar in its appearance ; it is exactly like other stars seen in nebulæ ; still it may really be but the brilliant condensed centre of a very remote cluster. I have not, however, detected any gradual increase of brilliancy towards the centre.

" Not to multiply sketches, which may soon require correction, I shall merely add, that in figure 32 we also find a star in the centre, and in figure 85 likewise a star in the centre, and many other minute stars in and close to it, so that it is really a cluster. The double nebulæ (figure 72) consists of two clusters, between which there is a star easily seen on even an indifferent night. In figure 49, there are minute stars between and about the three large stars, and I think there can be no doubt it is a cluster. Figure 25 abounds in stars, mixed with nebulosity ; I have not seen it on a very fine night, but it was observed by my assistant, and by a gentleman who was with him, and they had no doubt but that the centre was completely resolved. In the little annular nebulæ (figure 48) I see nothing remarkable, further than a star in the north preceding edge ; it is tolerably conspicuous, and is about half-way between the exterior and interior circumference of the annulus.

" Fɪɢ. 45 is a very remarkable object. It is no longer a planetary nebulæ, but an annular nebula, like that of Lyra, with a similarly fringed edge, though much less distinctly seen ; it is oval, but the central portion is not so dark as that of Lyra ; it very closely resembles the annular nebula of Lyra, seen with an instrument of inferior power."—*Observations,* pp. 2-3.

The accompanying plate (IV.) represents two of these interesting objects as shown in the three-feet reflector, selected from among five which illustrate his Lordship's paper in the *Philosophical Transactions.*

The first of these (numbered 81 in Herschel's catalogue and 51 in that of Messier), had hitherto been represented as of an *oval* figure. A glance at the plate will enable the reader to judge of the power of Lord Rosse's instrument. The singularly fantastic filaments which shoot from its southern extremity, have been detected by the superior illuminating power of the new telescope. They change the entire character of the figure ; and it is not improbably conjectured by Lord Rosse, that a further increase of power might bring out other filaments, and thus restore it to the ordinary form of a cluster.

The second is a much more splendid object. It is numbered 26 in Herschel's catalogue, and from its supposed form has hitherto been called the *Dumb-bell Nebula*. Our plate, however, will show how far this name is from giving a true idea of the shape and extent of the cluster. It is found to want the exact elliptical termination of the figure under which it was shown in Herschel's telescope, and to be of infinitely greater extent, and entirely different character. Another of Herschel's nebulæ (numbered 45 in his catalogue), which he represented as a *planetary* nebula of a *round* figure, is shown by Lord Rosse to be of an *annular* form, like that of Lyra, though much less distinctly seen.

From these comparisons it will be seen, not only, in general, that the observations hitherto made upon nebulæ are too imperfect to form a safe foundation for any hypothesis; but also, that, in particular, the determinate nebulæ pointed out by Herschel as globular in form, are not really globular, but of most irregular and unsymmetrical figures.

Lord Rosse's observations extended to about two-thirds of the figured nebulæ; and in many others he discovered other less remarkable discrepancies of form from that assigned in Herschel's figures.* On the general question of the ultimate resolvability of all nebulæ, he offers no decisive opinion.

" We should err were we to assume, that the absence of resolvability was evidence conclusive that the object was not a cluster. In some instances, with increasing optical power, the resolvable character has become clearly developed (as in figure 26), and a further increase of power has shown the object resolved. It is also important to observe, that now, as has always been the case, an increase of instrumental power has added to the number of clusters, at the expense of the nebulæ properly so called; still it would be very unsafe to conclude, that such will always be the case, and thence to draw the obvious inference, that all nebulosity is but the glare of stars too remote to be separated by the utmost power of our instruments."

But we must draw to a close. And, indeed, there are few who may not shrink at the wild and dreamy speculations which this startling subject suggests. How elevating, yet how humbling, the conceptions it forces upon the mind! How it overpowers us with the consciousness of the limitlessness of the works of the Great Creator! Take the nebula of Orion as an example. Suppose its distance from

---

* We shall anxiously look for his Lordship's report upon the Moon, and upon the general character of Messrs. Bäer and Maedler's map of the lunar surface He states, that wherever the three-feet reflector has been tried, it has already brought out many new details not noticed by them.

our earth to be only that of a star of the eighth magnitude. Even if this be its real distance from us, a portion of it, only 10′ in diameter, must spread through a space exceeding the dimensions of our sun, more than 2,000,000,000,000,000,000,000 times! And yet this is but an atom in the masses of nebulous matter which we see all around!

Now, to realize in some faint degree the infinite immensity of the universe, which this and similar observations necessarily imply, let it be only imagined (what indeed the telescope demonstratively shows for most of them), that each of these distant masses is a system, like the vast system in which ours is but a small unit—a system as extensive, as glorious, as perfect as our own; but so immensely distant, that the rays of the single bodies which compose it—its suns and stars—are blent together in their passage towards us, and, as it were, interpenetrate each other, so as to appear to our eyes but one luminous mass. And, to give the full grandeur of this conception somewhat of a palpable form, let it be conceived, that our whole visible universe, with all its parts, its myriads of suns and stars, which we see all around us, *is but a nebula in the vastness of space!* that, just as the nebulæ appear nebulous to us, so also to an observer in one of the worlds of some of these distant regions, all the bright and glorious orbs we see, not alone our own little system—the sun which rules our day, and the moon which lights up our night—but sun and stars and all—Sirius, and Procyon, and Capella, stars of the first, and stars of the twelfth magnitude—bodies separated from each other by myriads of millions of miles—are all brought together into such apparent proximity, as to be seen under an almost imperceptible angle, and in the appearance of a faint patch of misty light! Above all, let it be imagined—if indeed the mind can bear this stretching of its powers—that, as we see these wondrous bodies in every quarter of our heavens, so to an observer in the most remote of them all, at the farthest point of space to which the most powerful telescope can ever hope to penetrate, similar appearances may present themselves, not alone in the direction of our system, or in those directions to which our sight can reach, but to the north, and the south, and the east, and the west;—let all this be imagined, and *perhaps* (for, after all, these are but *possible* limits), some idea may then be formed of the force of the Lord's address to the Patriarch, *Suspice cœlum, et numera stellas* SI POTES!

APPENDIX.

Since the publication of this paper the performance of the six-feet reflector has been tested by the Earl of Rosse, assisted by the Rev. Dr. Robinson of the Armagh Observatory, and Sir James South. A full and authentic account of these observations will no doubt soon be laid before the public. Meanwhile it may not be uninteresting to transcribe the following extract from a letter of Sir James South, which appeared in the Times Newspaper of April the 16th.

"THE night of the 5th of March was, I think, the finest I ever saw in Ireland. Many nebulæ were observed by Lord Rosse, Dr. Robinson, and myself. Most of them were, for the first time since their creation, seen by us as groups or clusters of stars; whilst some, at least to my eyes, showed no such resolution. Never, however, in my life did I see such glorious sidereal pictures as this instrument afforded us. Most of the nebulæ we saw I certainly have observed with my own large achromatic; but although that instrument, as far as relates to magnifying power, is probably inferior to no one in existence, yet to compare these nebulæ, as seen with it and the 6-feet telescope, is like comparing, as seen with the naked eye, the dinginess of the planet Saturn to the brilliancy of Venus.

"The most popularly known nebulæ observed this night were the ring nebulæ in the Canes Venatici, or the 51st of Messier's catalogue, which was resolved into stars with a magnifying power of 548; and the 94th of Messier, which is in the same constellation, and which was resolved into a large globular cluster of stars, not much unlike the well-known cluster in Hercules, called also 13th Messier.

"Although, however, the power of this telescope in resolving nebulæ into stars hitherto considered irresolvable was extremely gratifying, still it was in my mind little more than I had anticipated; for experience has long since told me that a telescope may show nebulæ, even those resolvable by it, very well, whilst, when directed to a bright star, with a very moderate magnifying power, its imperfections will be actually offensive. During Sir W. Herschel's lifetime, with the twenty-feet reflector at Slough I saw, amongst others, 3 Messier, 5 Messier, 13 Messier, 92 Messier, the annular nebula of Lyra, and the great nebula of Andromeda. No telescope of its size probably ever showed them better; yet on the same night the same instrument, when directed to Alpha Lyræ (a star of the first magnitude), broke down under a power of about 300.

*   *   *   *   *   *   *   *

"Regulus on the 11th being near the meridian, I placed the six-feet telescope on it, and with the entire aperture and a magnifying power of 800 I saw, with inexpressible delight, the star free from wings, tails, or optical appendages; not indeed like a planetary disk, as in my large achromatic, but as a round image resembling voltaic light between charcoal points; and so little aberrations had this brilliant image that I could have measured its distance from, and position with, any of the stars in the field with a spider's line micrometer, and a power of 1,000, without the slightest difficulty; for not only was the large star round, but the telescope, although in the open air and the wind blowing rather fresh, was as steady as a rock.

" On subsequent nights, observations of other nebulæ, amounting to some thirty, or more, removed most of them from the list of nebulæ, where they had long figured, to that of clusters; whilst some of these latter, but more especially 5 Messier, exhibited a sidereal picture in the telescope such as man before had never seen, and which for its magnificence baffles all description.

" Several double stars were seen with various apertures of the telescope, and with powers between 360 and 800; and, as the Earl had told us before we should,—before the speculum was inserted in the tube, in consequence of his having been obliged to quit the superintendence of the polishing at the most critical part of the process,—we found that a ring of about six inches broad, reckoning from the circumference of the speculum, was not perfectly polished; and to *that* the little irradiation seen about Regulus was unquestionably referable.

" The only double stars of the first class which the weather permitted us to examine with it were Xi Ursæ Majoris and Gamma Virginis; those I could have measured with the greatest confidence, whether however it would have separated some of the closest or of the most difficult double stars I cannot say.

"D'Arrest's comet we observed on the twelfth of March, with a power of 400, but nothing worthy of notice was detected.

"Of the moon a few words must suffice. Its appearance in my large achromatic, of 12 inches aperture, is known to hundreds of your readers; let them then imagine that with it they look *at* the moon, whilst with Lord Rosse's 6 feet they look *into* it, and they will not form a very erroneous opinion of the performance of the Leviathan."

.

Printed in the United States
By Bookmasters